CONTENTS

3   Continuing the Species   96

4   The Food Chain   124

1
Patterns of Life

All of the earth's creatures live in a biosphere made up of air, water, and land. Within this life zone, several million species of animals and about 500,000 species of plants interact in a continuing exchange that has shaped and reshaped the planet over some three billion years. The study of this interplay is known as ecology—from the Greek meaning "the study of the home."

Modern-day ecology has its roots in the nineteenth-century discovery that the world's life forms inhabit six distinct geographic regions: the Nearctic Region of North America; the Pale-

arctic Region of Eurasia and northern Africa; the Oriental Region of India, Indochina, western Indonesia, and the Philippines; the Neotropical Region of South and Central America, tropical Mexico, and the Caribbean islands; the Ethiopian Region of Africa, southern Arabia, and Madagascar; and the Australian Region of Australia, New Zealand, and the South Pacific islands.

Each region contains plant and animal species unlike any other. This is due in part to the earth's geological past and in part to the planet's varied climate zones and the regions, or biomes, they

control. The distribution of the species also owes much to evolution, the way in which life forms adapt to their homes. Chapter 1 will investigate how these factors have shaped today's rich and distinctive life patterns.

Earth's diverse forest biomes include, from top left, the boreal forest, the temperate rain forest, the temperate deciduous forest, and the tropical rain forest. Below, from left, are animals unique to the Ethiopian Region, the Oriental Region, the Australian Region, and the Neotropical Region.

Why Are Some Animals Unique to the Galápagos?

Far out in the equatorial Pacific, some 600 miles west of Ecuador, lie the volcanic islands known as the Galápagos. Hundreds of animal species live there, many unique to the Galápagos—among them 600-pound tortoises, flightless cormorants, and deep-diving marine iguanas.

Two factors contributed to the creation and preservation of this rich environment: the islands' extreme isolation and the area's odd climate (*right*). Long ago, some parent species arrived on natural rafts from South America to colonize the isles. Over time, these secluded animals evolved peculiar characteristics suited to their island home—and nowhere else in the world.

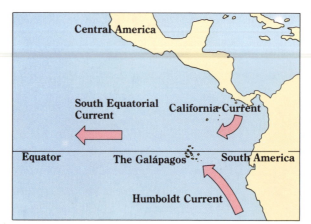

Central America

South Equatorial Current

California Current

Equator

The Galápagos

South America

Humboldt Current

Though they straddle the equator, the Galápagos Islands enjoy a temperate climate. The South Equatorial Current— a chilly oceanic river carrying waters from the Antarctic-fed Humboldt Current— and the California Current, fed by frigid North Pacific waters, cool the archipelago.

2. Giant tortoise. Two centuries of slaughter by humans has cut the Galápagos tortoise population from 250,000 to 15,000. Undisturbed, the animals live more than 100 years.

3. Swallow-tailed gull. The gulls breed only on the lava cliffs but sometimes fly to South America.

4. Frigate bird. The pirates of the Galápagos, two species of frigate bird steal other birds' prey in midair. In breeding season, males display red throat sacs to attract mates.

5. Galápagos penguin. These 20-inch-tall relatives of the cold-water penguins thrive in the cool waters of the tropical archipelago.

1. Land iguana. Though this fearsome-looking, 2-foot lizard hisses and spits when threatened, it is harmless. Fleshy cactus pads—spines and all—are its favorite food.

Darwin's finches

When British naturalist Charles Darwin visited the Galápagos Islands in September 1835, he discovered 13 species of finches. Though alike in size and plumage, each had a different beak. Darwin dismissed the idea that so many finch varieties could have roosted on the Galápagos. He theorized that the descendants of one original species had adapted to the archipelago's specialized habitats, over time becoming seed-eaters or insect-eaters—each with a beak suited to its needs.

Large ground finch; hard seed-eater

Warbler finch; insect-eater

Woodpecker finch; insect-eater, tool-user

Large tree finch; insect- and seed-eater

Sharp-beaked ground finch; seeds, insects

Large cactus ground finch; cactus-eater

6. Marine iguana. The islands' seagoing lizards feed on kelp and other seaweed, diving to 40 feet and staying as long as 30 minutes.

7. Galápagos sea lion. Some 50,000 sea lions, a subspecies of the California sea lion, inhabit the Galápagos. Streamlined bodies make them swift and graceful swimmers.

8. Flightless cormorant. Galápagos cormorants use their wings for diving and for balance on land. The tiny wings and huge feet may also function as heat regulators.

Do Penguins Live at the North Pole?

▶ **Murre.** The largest of the 22 existing species of auks, the adult weighs 2 pounds and is just over a foot tall. Each year, the females lay one egg apiece on the cliffs.

At first glance, the black-suited seabirds that live in the Northern Hemisphere's Arctic *(above)* and more temperate regions could be mistaken for penguins: They inhabit a chilly marine environment, and their densely feathered bodies, short wings, and webbed feet make them superb divers and swimmers. But these northern birds are auks, which nest on high cliffs or in burrows beyond the reach of predators.

Penguins evolved in the oceans of the Southern Hemisphere surrounding Antarctica. Safe from land-based predators, the penguins adapted to swimming and have flipperlike wings. But they could never reach the North Pole, for they would have to cross the warm seas of the tropics.

An auk sampler

Auks *(below)* live only in the Northern Hemisphere.

The company-loving razorbills breed alongside murres on North Atlantic cliffs.

Using its beak, the horned puffin often digs a shallow burrow in which to hide its egg.

The ancient murrelet, named for its gray head feathers, winters on the open sea.

The now-extinct great auk was the largest and only flightless member of its family.

Emperor penguin. The nearly 4-foot-tall, 70-pound emperor breeds on the ice shelves of Antarctica. It can remain submerged for 18 minutes while scouring the ocean depths for squid and fish.

▼**All 18 penguin species** reside in the Southern Hemisphere. But only two—the emperor and the Adélie— brave the extreme cold of Antarctica.

▲ **Powered by** oarlike flippers, the Adélie penguin swims at a speed of 22 miles per hour.

South America

Africa

+ South Pole

Antarctica

Australia

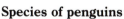

Species of penguins

Named for the marking under its throat, the chinstrap penguin prefers rocky slopes.

The king penguin, smaller than the emperor, breeds in warmer, subantarctic islands.

Coastal South American natives, Magellanic penguins breed in adjoining burrows.

The agile rockhopper penguin navigates the rocky ground with spirited hops.

Why Are the Animals of India and Africa So Alike?

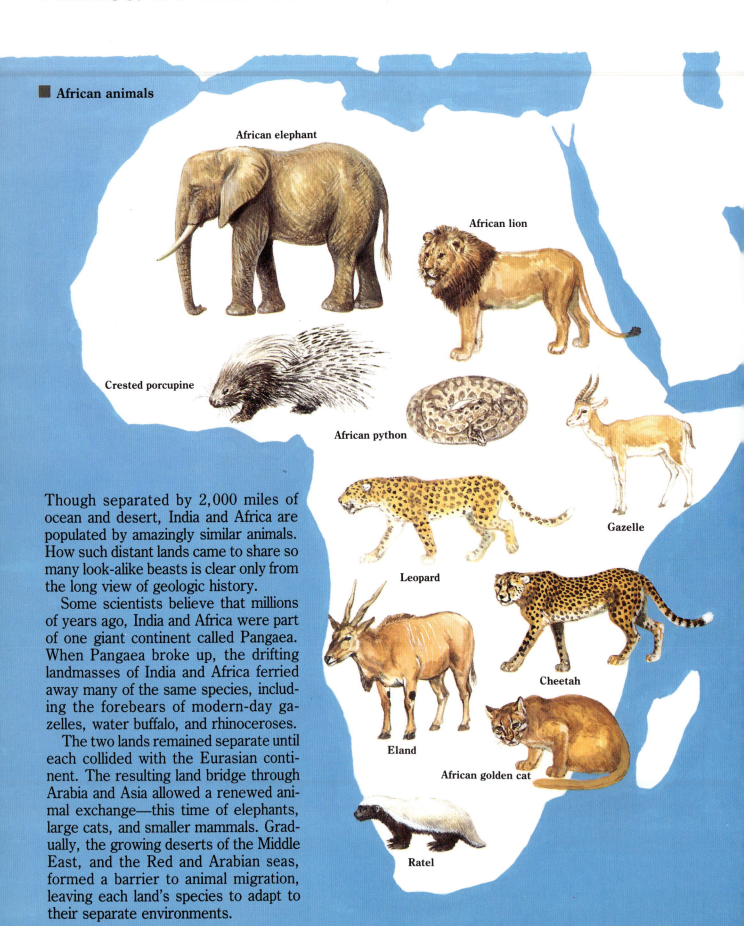

■ African animals

African elephant

African lion

Crested porcupine

African python

Gazelle

Leopard

Cheetah

Eland

African golden cat

Ratel

Though separated by 2,000 miles of ocean and desert, India and Africa are populated by amazingly similar animals. How such distant lands came to share so many look-alike beasts is clear only from the long view of geologic history.

Some scientists believe that millions of years ago, India and Africa were part of one giant continent called Pangaea. When Pangaea broke up, the drifting landmasses of India and Africa ferried away many of the same species, including the forebears of modern-day gazelles, water buffalo, and rhinoceroses.

The two lands remained separate until each collided with the Eurasian continent. The resulting land bridge through Arabia and Asia allowed a renewed animal exchange—this time of elephants, large cats, and smaller mammals. Gradually, the growing deserts of the Middle East, and the Red and Arabian seas, formed a barrier to animal migration, leaving each land's species to adapt to their separate environments.

Indian animals

Indian elephant

Indian lion

Indian python

Indian gazelle

Indian porcupine

Leopard

Indian cheetah (extinct)

Indian blue bull

Ratel

Indian golden cat

Relatives on restless lands

The similarities among the animals inhabiting the earth's far-flung continents are no coincidence. About 265 million years ago, powerful geologic forces caused the planet's landmasses to gather into one supercontinent known as Pangaea. Ranging across this expanse were the distant ancestors of today's animal kingdom.

When the landmasses of Pangaea began to drift apart millions of years later, forming the seven continents of today's world, primitive reptile and mammal species were carried along with them. Over time, climatic changes on the continents caused these creatures to evolve differently. However, remarkable commonalities persist among species today: The mammals and reptiles of India and Africa are one example; the flightless birds that populate New Zealand, Australia, Africa, and South America are another.

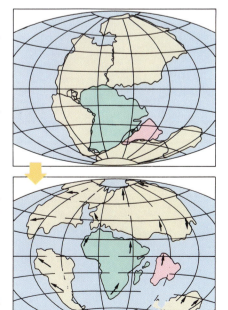

All the earth's land surface was once part of the supercontinent called Pangaea. Within this giant landmass, India *(pink),* then a separate land fragment, nestled next to Africa *(green).* Animals living in the two regions roamed freely back and forth.

When Pangaea broke up, the island continent of India *(pink)* split away from Africa *(green)* and began floating northward. Like a giant ark, it carried along the animals common to its one-time neighbor.

How Do Reptiles Survive in Cold Climates?

Life of the garter snake

Male garter snakes, sluggish after a long winter's hibernation, emerge from their underground den onto sun-warmed rocks.

Entwining themselves around a newly awakened female *(large snake, center),* male snakes numbering from 10 to 100 form a "mating ball."

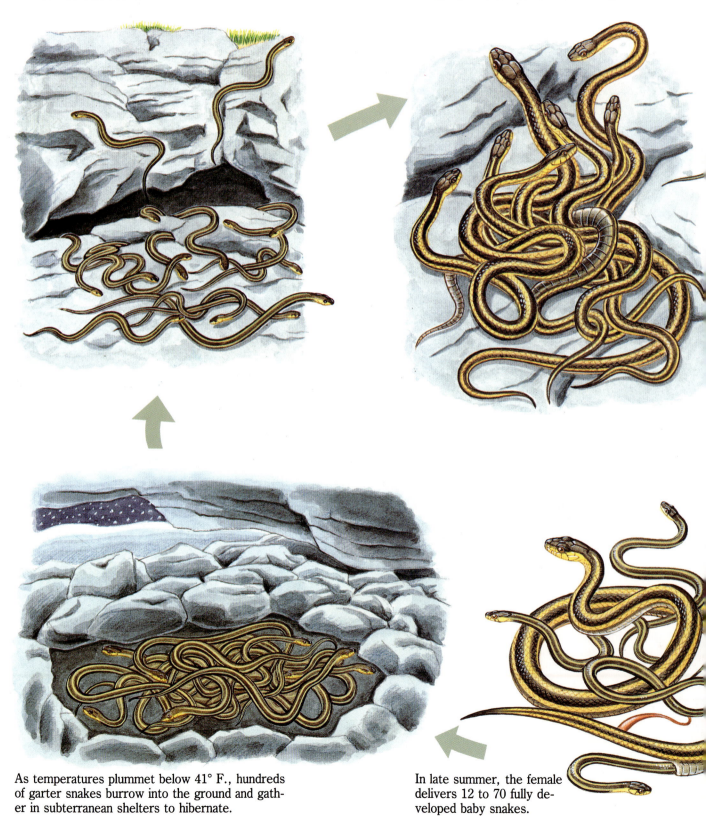

As temperatures plummet below 41° F., hundreds of garter snakes burrow into the ground and gather in subterranean shelters to hibernate.

In late summer, the female delivers 12 to 70 fully developed baby snakes.

Unlike warm-blooded mammals and birds that maintain a constant internal body temperature, reptiles are cold-blooded creatures that absorb heat from outside sources. For this reason, most reptiles—including snakes, lizards, turtles, and crocodiles—inhabit tropical regions. A few, however, have adapted to surprisingly unreptilian environments. Two species—a lacertid lizard and the common viper—thrive above the Arctic Circle in Europe. The garter snake *(below),* the most common snake in North America, dwells as far north as the frosty Yukon of northwest Canada.

These reptiles are so efficient at siphoning warmth from their environs that they can raise their body temperatures by as much as 50° F. over the surrounding air temperature by sunbathing. When the weather becomes too cold for the creatures to function normally, they fall into a deep sleep known as hibernation, in some cases for as long as nine months.

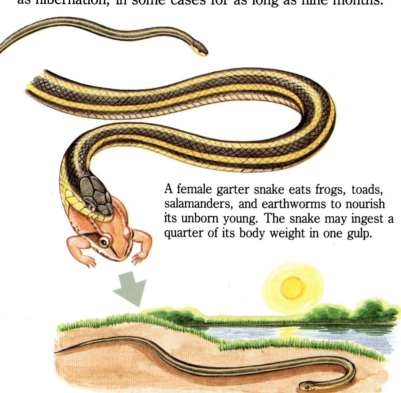

A female garter snake eats frogs, toads, salamanders, and earthworms to nourish its unborn young. The snake may ingest a quarter of its body weight in one gulp.

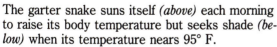

The garter snake suns itself *(above)* each morning to raise its body temperature but seeks shade *(below)* when its temperature nears 95° F.

● Where reptiles reside

Most of the world's 6,000 reptile species live within the region bordering the earth's equator. Above latitudes 40° N and 40° S—from the mid-United States, Eurasia, and Japan north, and from lower Argentina and New Zealand south—the reptile populations drop off rapidly. The maps below show the number of reptile species living in Europe and Japan.

Reptile species in Europe

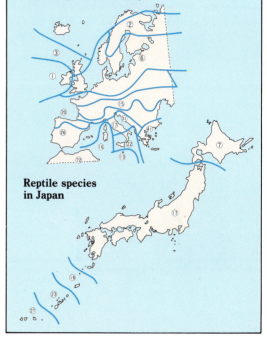

Reptile species in Japan

A cold-blooded advantage

Compared to mammals, which rapidly convert the food they eat into body heat, reptiles metabolize their food slowly, generating almost no internal warmth. Cold-blooded creatures have an advantage, however, when it comes to foraging: Reptiles require far less food than their warm-blooded competitors. A 50-pound African hunting dog, for example, needs 7 pounds of meat a day, whereas a 300-pound Komodo dragon gets by on 1 pound.

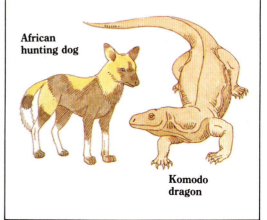

African hunting dog

Komodo dragon

Why Are Tapirs Found Only in South America and Asia?

The tapir, a distant relative of the horse and the rhinoceros, is one of the most primitive mammals on earth. An animal that resembles a cross between a pig and an elephant, the tapir has changed little during its 55 million years on earth.

Though threatened with extinction, these ancient forest dwellers still survive in regions far removed from their original home. Some 50 million years ago, ancestors of the modern-day tapir migrated from their native North America into Asia and Europe, and south into Central and South America. Climate changes pushed their realm to the tropics. Today their habitat is limited to the tropical rain forests of Southeast Asia and a range in the Americas that stretches from southern Mexico to Brazil.

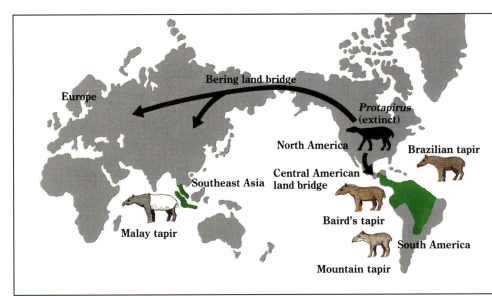

Europe

Bering land bridge

Southeast Asia

Malay tapir

North America

Protapirus (extinct)

Central American land bridge

Baird's tapir

Mountain tapir

Brazilian tapir

South America

Tapirs' progress

Fossils suggest that tapirs first appeared in North America. Tapirs spread from there into Asia and Europe across the Bering land bridge that once connected the continents. With the onset of colder temperatures in northern latitudes, however, the European tapirs died off and the Asian species dwindled. Their North American counterparts headed south, over the new Central American land bridge and into South America.

Malay tapir

Sometimes referred to as the blanket tapir for the white sheathlike marking on its back, this tapir roams the humid forests of southern Burma and Thailand, the Malay Peninsula, and Sumatra, Indonesia. It is the only species of tapir remaining in Asia.

● **Mountain tapir**

These smallest of tapirs are protected against the cold of their Andean mountain habitat by a dense coat of inch-long brown hair. They are found throughout central Colombia, Ecuador, and northern Peru.

● **Brazilian tapir**

The widest-ranging species is the Brazilian tapir, an inhabitant of Venezuela, Brazil, Paraguay, and northern Argentina. In settled areas, Brazilian tapirs invade plantations at night to feast on sugarcane.

● **Baird's tapir**

The giant among tapirs, Baird's tapir weighs nearly 700 pounds. This species dwells in mangrove swamps, tropical forests, woodlands, and mountain terrain from Mexico south to Panama.

Why Do Marsupials Thrive in Australia?

Only in Australia do the mammals known as marsupials dominate the landscape. Best known for carrying their developing young in a marsupium, or abdominal pouch, these strange creatures fill niches commonly occupied elsewhere by placental mammals, which nourish their offspring in the womb. Thus it is kangaroos, not gazelles, that forage the Australian grasslands, and koalas, not monkeys, that live in the trees.

Marsupials developed millions of years ago on a continent that included South America, Africa, Australia, and Antarctica. Later, when the continent broke apart, the South American landmass connected with North America. Most South American marsupials were replaced by the more advanced placental mammals that evolved in the north and moved south, leaving most of the marsupials in Australia.

March of the marsupials

When the earliest mammals evolved, about 265 million years ago, the earth's landmass consisted of one supercontinent called Pangaea.

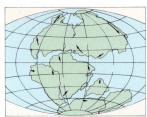

About 140 million years later, Pangaea split into two major continents. The first marsupials developed on the South American landmass in the great southern continent.

Over the next 30 million years, as the continents began to divide further, the marsupials traveled south to the landmass that is Australia today.

Marsupials by habitat	
Habitat	**Species of marsupials**
Subsurface	Mole
Surface	Bandicoot, wombat, anteater
Grasslands	Kangaroo, wallaby
Trees	Sugar glider, striped possum, cuscus, koala, marsupial cat

Like their placental counterparts, Australian marsupials have adapted and flourished in a variety of habitats.

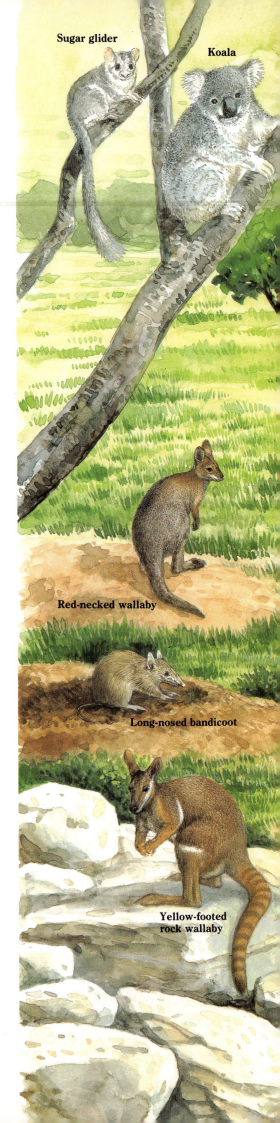

Sugar glider

Koala

Red-necked wallaby

Long-nosed bandicoot

Yellow-footed rock wallaby

■ Australia's varied marsupials

Striped possum

Black tree kangaroo

Gray cuscus

Spotted cuscus

Eastern gray kangaroo

Red kangaroo

Marsupial cat

Parma wallaby

Common brush-tailed possum

Banded anteater

Common wombat

Marsupial mole

17

What Became of North America's Monkeys?

The only monkeys in North America today live in zoos. But in the remote past, for a period spanning 30 million years, the prosimian *Plesiadapis rex, Adapis,* and *Omomyidae*—the primate ancestors of today's monkeys—wandered across a land bridge from Europe and Asia to inhabit much of the North American continent. About 20 million years ago, a cooler, drier climate descended on the Northern Hemisphere, bringing with it a great ice age that killed off the primates in North America and Europe.

The bitter cold of the ice age did not extend into South America, Africa, or southern Asia, sparing the primate populations in those regions. When the northern glaciers receded, these early primates were not able to recolonize the cold regions. And South American monkeys, isolated until some three million years ago when the northern and southern continents joined up, found the journey north too dry.

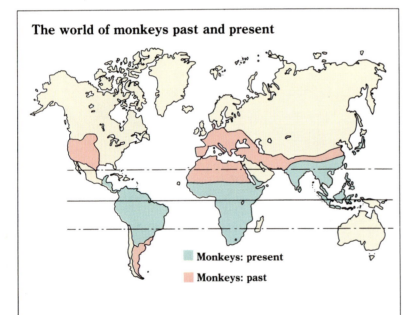

The world of monkeys past and present

Monkeys: present

Monkeys: past

Today, monkeys only inhabit the earth's tropical and subtropical regions *(blue)*. About 55 million years ago, when the planet was warmer, monkeys could be found throughout large portions of North America, southern South America, Europe, northern Asia, and Africa *(pink)*.

Picrodus sillberlingi

■ Extinct primate precursors of North America and Europe

Pronycticebus

Europolemur

Chiromyoides minor

Microsyops

19

How Do the Animals of South America Compare to Those of Africa?

South America's giant rodents and certain other orders of mammals are the natural consequence of one of evolution's more bizarre ground rules: Unrelated species inhabiting similar environments gradually evolve similar habits and appearances. Millions of years ago, when the supercontinent of Pangaea dominated the globe, the first rodents appeared. When the South American landmass broke away millions of years later, it

carried with it many rodents from Pangaea.

Meanwhile, a new order of animals was evolving on the still-linked North American and Eurasian continents. These were artiodactyls—hoofed ancestors of the giraffe, antelope, and buffalo. Artiodactyls flourished in their homeland, taking over new habitats.

When continental drift brought the African-Arabian landmass into contact with the Eurasian

Animals of South America

▶ Megathere, an extinct giant sloth, nibbled leaves like a giraffe.

▲ The squirrel-like spiny tree rat forages for nuts in the treetops.

The paca inhabits forests from central Mexico to southern Brazil. It measures 30 inches long and weighs 20 pounds. ▼

The capybara's foot, ▶ like the pygmy hippo's, has broad toes suitable for support on swampy ground.

▲ The world's largest rodent, the 100-pound capybara feeds on marsh grasses and other kinds of aquatic vegetation.

◀ The mara has the short forelegs and longer hind legs of the African duiker.

◀ Equipped with rabbit-like haunches for running, the agouti lives in a burrow.

The mara's feet, like ▶ the African duiker's hoofs, are adapted for the bush.

20

Millions of years ago, hoofed animals spread through Eurasia and North America. On the island landmass of South America, rodents developed instead.

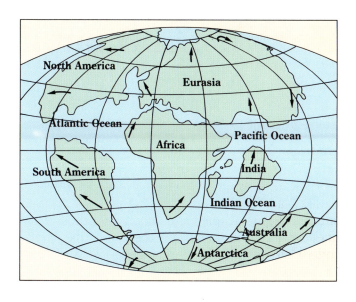

continent, hoofed animals moved into Africa. They did not reach South America, however, until it linked up with North America three million years ago. Thus South America's rodents, lacking competitors for many millennia, filled some of the ecological niches occupied by hoofed animals in North America, Eurasia, and Africa. As a result, the forms and behaviors of the ecological analogous orders became strangely similar.

Animals of Africa

The Congo tree mouse, a member of the rodent family, is a tree dweller.

Leafy treetops are the feeding ground of choice for the 12-foot-tall giraffe.

Like the capybara, the pygmy hippopotamus frequents wetland habitats.

The pygmy hippo's splayed toes provide sure footing on boggy ground.

The musk dwarf antelope, which has large musk glands below the eyes, browses the forests of central Africa.

Hoofs sheathe the common duiker's toes, lending them stability.

Found in the grass-lands of northern Africa, the duiker is fast and agile.

Smith's red rock hare, similar in size to the agouti, eats grasses.

21

How Are Coconut Palms Spread?

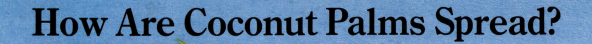

Common along tropical coasts throughout the world is a tall, unbranched tree called the coconut palm. Although the origin of the tree is unknown, it has grown in southern Asia and the Malay islands since prehistoric times. Its fibrous trunk sprouts a tuft of leaves measuring up to 20 feet long and a cluster of coconuts.

The coconut palm multiplies by sending out "seed boats"—ripened coconuts that fall from seaside palms into coastal waters and are swept along by ocean currents to new lands. The foot-long coconuts stay afloat because their spongy outer shell is shot through with air pockets. If a coconut makes landfall in a region with abundant sunshine, humidity, and rainfall, it will grow to a mature palm—about 50 feet tall—within eight years and will bear up to 100 coconuts annually.

Parts of the coconut

Germination pores

Epicarp (outer shell)

Endocarp (inner shell)

Endosperm (meat)

Mesocarp (middle layer)

Embryo

Germination begins when a ripe coconut washes ashore. The embryo, nourished by rainwater trapped in the corky mesocarp and by nutrients in the endosperm, grows through the fibrous outer shell. The seedling feeds off the fleshy endosperm until the fledgling palm is firmly rooted.

A coconut beached on rocky
▲ or stony soil will not take root.

If a coconut lands in the
▲ shade, the seedling will wither.

After three months on a
sunny, sandy beach, a coconut
seedling begins growing roots.

Why Do Forests Differ Worldwide?

From Brazil's lush rain forests to Canada's vast evergreen stands, forests are among the most complex and diverse biomes, or ecological communities, on the planet. Accounting for 30 percent of the earth's land surface, they arise only in regions rainy enough to support their substantial thirst. The types of trees—deciduous or evergreen—that grow in a given forest are determined by the region's geography and its climate.

Deciduous trees lose their leaves during cold or dry seasons, while evergreens remain green year-round. Either type of tree may be of the broad-leaved or needle-leaved variety. Regions with balmy summers and cold winters promote the growth of deciduous broad-leaved forests, where trees shed their leaves with the first frost, then bud again in the spring. In the snowbound northern regions, needle-leaved evergreens abound. The fine, waxy leaves of these trees keep moisture loss to a minimum. Broad-leaved evergreens thrive in warm, wet climates where their leaves collect the sunshine and moisture.

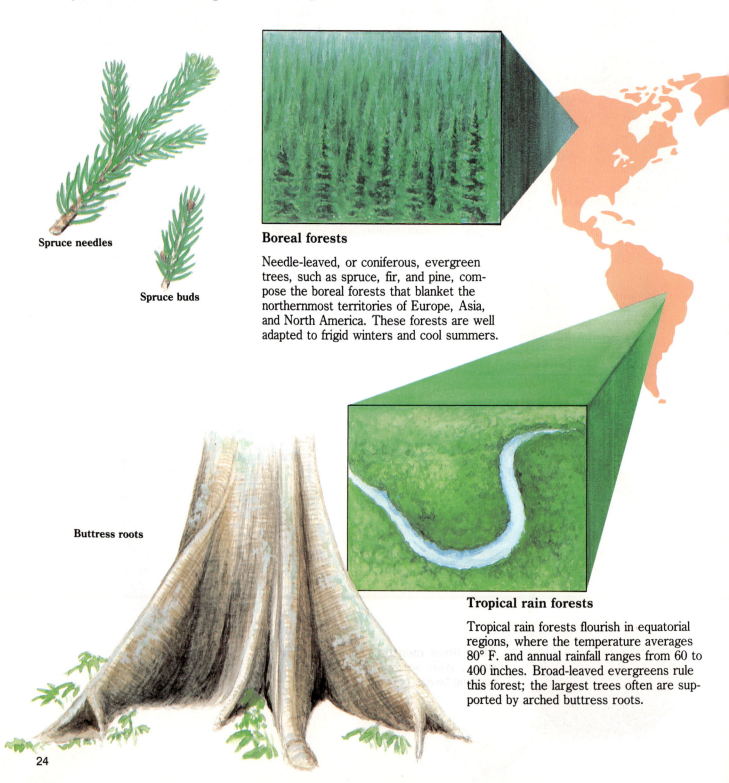

Spruce needles

Spruce buds

Boreal forests

Needle-leaved, or coniferous, evergreen trees, such as spruce, fir, and pine, compose the boreal forests that blanket the northernmost territories of Europe, Asia, and North America. These forests are well adapted to frigid winters and cool summers.

Buttress roots

Tropical rain forests

Tropical rain forests flourish in equatorial regions, where the temperature averages 80° F. and annual rainfall ranges from 60 to 400 inches. Broad-leaved evergreens rule this forest; the largest trees often are supported by arched buttress roots.

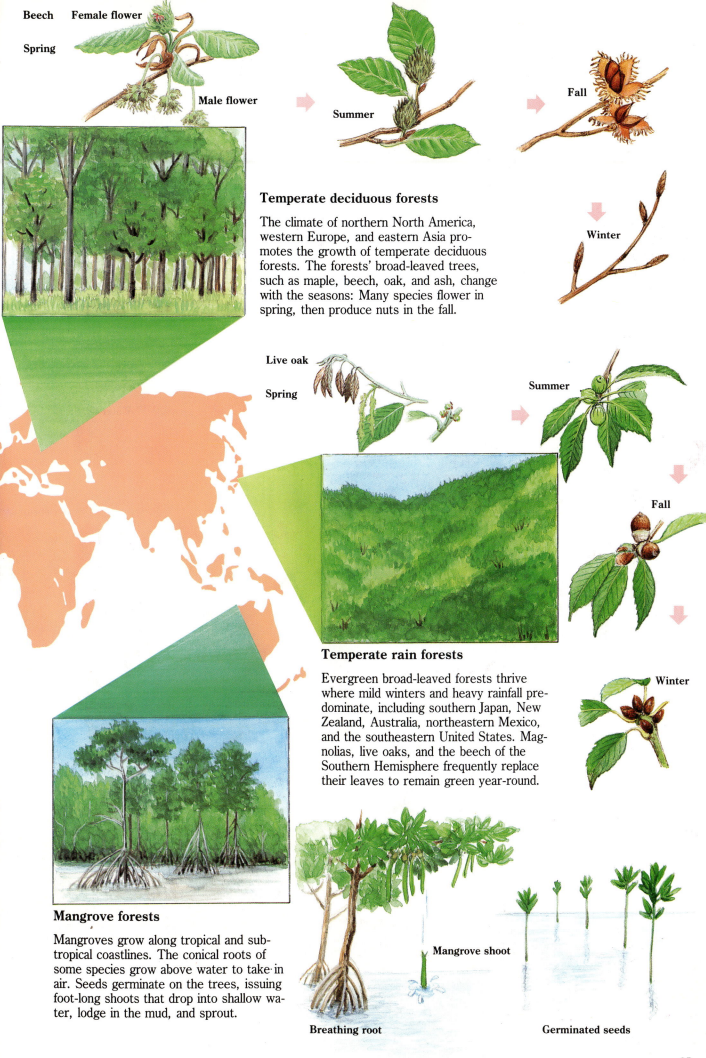

Beech Female flower

Spring

Male flower

Summer

Fall

Winter

Temperate deciduous forests

The climate of northern North America, western Europe, and eastern Asia promotes the growth of temperate deciduous forests. The forests' broad-leaved trees, such as maple, beech, oak, and ash, change with the seasons: Many species flower in spring, then produce nuts in the fall.

Live oak

Spring

Summer

Fall

Winter

Temperate rain forests

Evergreen broad-leaved forests thrive where mild winters and heavy rainfall predominate, including southern Japan, New Zealand, Australia, northeastern Mexico, and the southeastern United States. Magnolias, live oaks, and the beech of the Southern Hemisphere frequently replace their leaves to remain green year-round.

Mangrove forests

Mangroves grow along tropical and subtropical coastlines. The conical roots of some species grow above water to take·in air. Seeds germinate on the trees, issuing foot-long shoots that drop into shallow water, lodge in the mud, and sprout.

Breathing root

Mangrove shoot

Germinated seeds

How Can Plants Live in Water?

Scientists theorize that plants first evolved on land, then gradually adapted themselves to life in the earth's lakes, rivers, and oceans. Some took to deep water, rooting themselves in muddy lake bottoms or silty seafloors. Others led a kind of double life on the shore—half in and half out of the water. A few abandoned the land altogether, becoming free-floating plants.

Each transition brought about fascinating changes in plant anatomy. To withstand the wear and tear of relentless currents, numerous water plants—also known as hydrophytes—developed slender, flexible stalks and ribbonlike leaves. Certain tiny, free-floating plants dispensed with their roots entirely, absorbing nutrients and water through their leaves instead. Land-rooted water plants grew stems with tiny air chambers inside that helped them stay afloat. These air-filled stems and leaves also helped to channel oxygen rootward.

Species of water plants

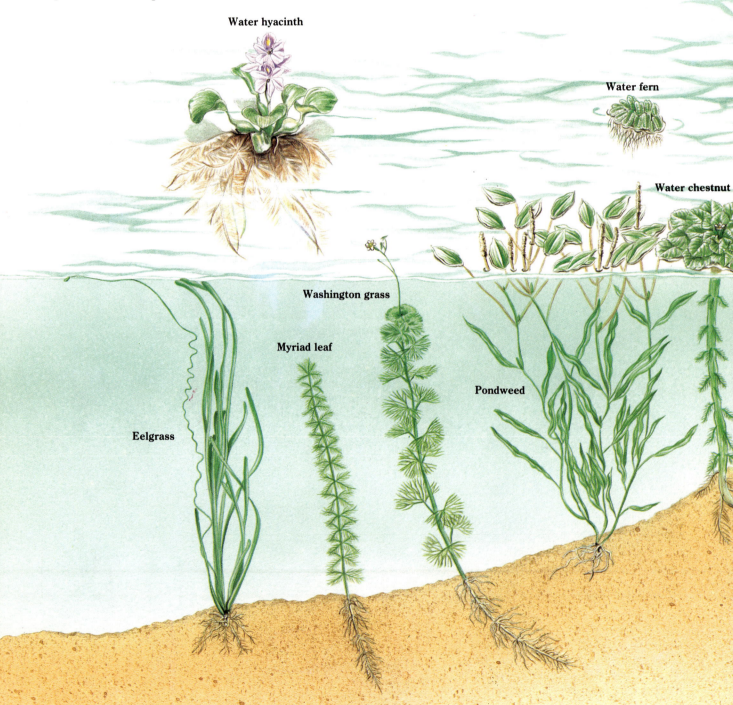

Water hyacinth

Water fern

Water chestnut

Washington grass

Myriad leaf

Pondweed

Eelgrass

Lotus

Japanese water lily

Ditch reed

● The mechanics of flotation

Frog's-bit

Leaf underside

Leaf surface

Flotation bladder

Stem cutaway

Hydrophytes, such as frog's-bit, that have leaves resting on the water's surface have an air-filled flotation bladder beneath each leaf. Like a tiny life jacket, the bladder keeps the leaf afloat. Air chambers in the stem (*cutaway*) keep the plant upright and supply oxygen to the roots.

Water plant pollination

Though most flowering aquatic plants rely on wind and insects to disperse their stores of pollen, a few, like the eelgrass pictured above, use water. Bobbing just below the water's surface, eelgrass flowers release tiny pollen grains into the passing current. The grains drift downstream until they come in contact with a female eelgrass plant, fertilizing it.

Types of water plants

Submersed hydrophytes grow underwater; free-floating plants suspend their top leaves on the water's surface; and emergents reach above water.

Iris

Japanese waterweed

Ditch reed

Mad-dog weed

Water oats

Japanese water lily

Frog's-bit

Pygmy water lily

Myriad leaf

Cattail

Eelgrass   Hydrilla

Pondweed

Water chestnut

Submersed plants ●

Free-floating plants ●

Emergent plants ●

K. MURAKAMI

2

Adapting to Life on Earth

Unlike any other planet in the solar system—and, perhaps, in the galaxy—earth has given rise to life forms of astonishing variety. Bathed in sunlight, the planet's atmosphere, oceans, and landmasses interact to produce global climate. At the same time, local patterns of rainfall, wind, and temperature yield landscapes with unique characteristics.

Together these forces have created nine major biomes, or zones, that share common communities of plant and animal life. Progressing roughly from the equator poleward, they are the

tropical rain forest; the savanna; the grassland; the hot, dry desert; the mild, shrubby chaparral; the temperate deciduous forest; the temperate coniferous forest; the cold taiga, or boreal forest; and the icy polar tundra.

The local environment in which a group of organisms lives is known as a habitat. Most species exist only in select habitats: Alpine plants, for example, grow on mountain slopes, while monkeys inhabit the canopy of the rain forest. Other species range from one habitat to another; salmon, for instance, migrate from the ocean to upriver spawning grounds. In each biome, certain plants and animals find niches—that is, specific opportunities for shelter or sustenance. The creatures best able to take advantage of these opportunities are the ones that survive—and thrive.

The biomes shown at top include, from left, the Arctic tundra, the desert, the boreal forest, and the savanna. The habitats below feature a cave occupied by bats and salamanders, a rain forest populated by monkeys and toucans, and an ocean that supports turtles and gulls.

K. MURAKAMI

What Determines a Turtle's Gender?

Hatchlings head for the sea

The sandy road to life

A female sea turtle digs a hole with its hind flippers. It lays 30 to 150 eggs and covers them with sand.

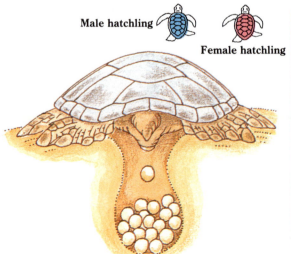

Male hatchling

Female hatchling

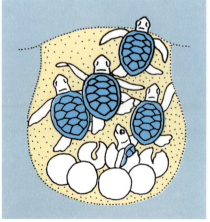

Turtle nests that never get warmer than 80° F. typically yield male hatchlings *(above)*.

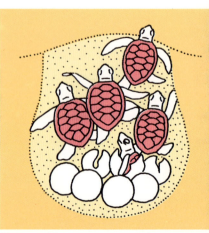

Nests with temperatures above 80° F. normally yield females.

Unlike most animals, whose sex is determined by chromosomes from a parent, sea turtles and some other reptiles owe their gender to an environmental factor: the amount of heat the turtle egg receives during incubation. If an egg is subjected to cooler temperatures during its one- to two-month sojourn in a sandy nest, the hatchling will emerge as a male; if the egg incubates at warmer temperatures, the hatchling will be a female. Eggs closest to the sunbaked surface generally produce female turtles. Those in the cooler bottom of the nest usually produce males.

Upon hatching, the turtles scurry down to the sea. (The brightness of the moon over the ocean and the downhill slope of the sand may help to orient them.) Although many are eaten by frigate birds and gulls, about 10 percent of each brood survives the race to the sea.

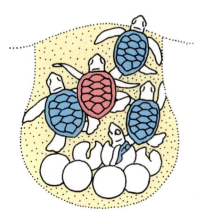

Clutches of turtle eggs laid in the cooler months of early spring hatch mostly males.

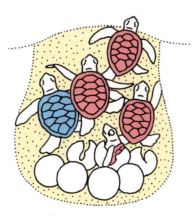

Eggs laid at the height of summer produce a preponderance of female sea turtles.

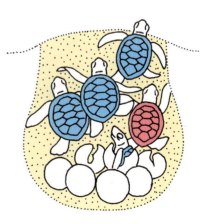

As temperatures drop in autumn, the male hatchlings again begin to outnumber the females.

Where Do Seabirds Live?

Although some seabirds fly hundreds of miles out to sea in search of plankton, squid, fish, and other food, most of them return to dry land to roost—that is, to rest for the night—and all of them return to nest, or raise their young. On land, more than 95 percent of seabirds (there are about 275 species in all) nest in colonies, which can range in size from just a few breeding pairs to more than a million birds. Most seabirds build their nests in sites designed to keep predators at bay: on rock outcroppings, beneath cliff overhangs, inside shallow caves, or atop promontories.

■ **A seabird sampler**

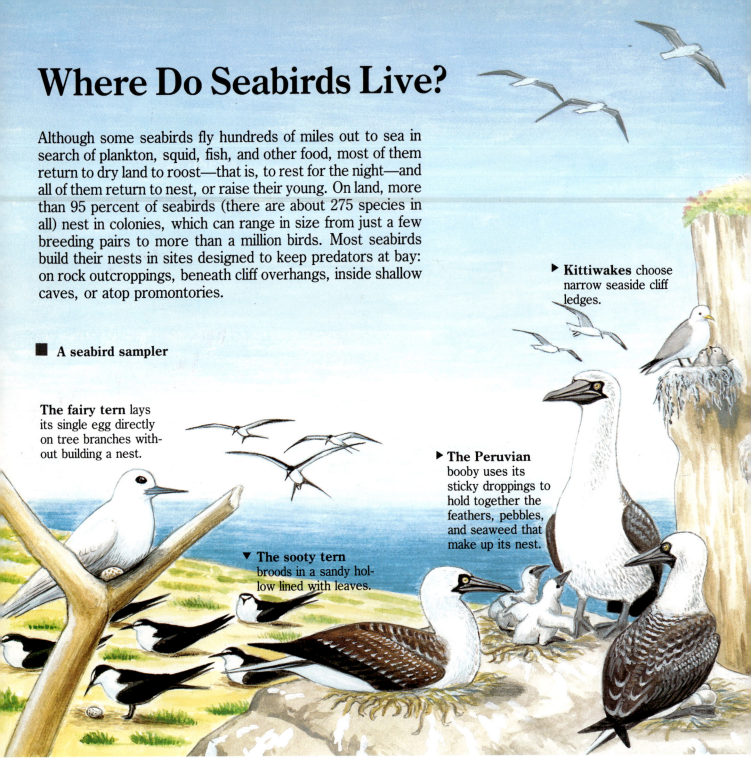

Kittiwakes choose narrow seaside cliff ledges.

The fairy tern lays its single egg directly on tree branches without building a nest.

▶ **The Peruvian** booby uses its sticky droppings to hold together the feathers, pebbles, and seaweed that make up its nest.

▼ **The sooty tern** broods in a sandy hollow lined with leaves.

Safety in numbers

In a congested living arrangement typical of seabirds around the world, northern gannets cluster in a dense colony *(right)* on a rocky island in the North Atlantic. Such group nesting confers several benefits: protection from predators, ease of finding a mate, and help in locating food.

▶ **Grass** and seaweed are the main materials for nests of the glaucous-winged gull.

▶ **The thick-billed** murre lays its egg on a sea-cliff ledge. The egg's shape keeps it from rolling off.

▼ **The tufted puffin** digs a burrow or finds a rock crevice, then lines it with feathers.

▼ **Seaweed,** grass, and droppings make up the nest of the pelagic cormorants.

▼ **A rock crevice** serves as a hatchery for ancient murrelets *(below).*

Return of the albatross

By the dawn of the twentieth century, feather hunters had all but eradicated the short-tailed albatross *(right)* from islands in the North Pacific. Between 1887 and 1903, Japanese plume hunters killed five million of the birds, whose feathers were used to stuff bedding and whose bones were made into pipestems. By the mid-1980s, however—thanks in part to a government program—the albatross had made a comeback on the island of Tori-shima.

How Do Mangrove Trees Survive in Salt Water?

Most trees are easily damaged or killed by salt water, but mangroves have evolved mechanisms that enable them to thrive in it. Not only do the trees' internal tissues exhibit a high tolerance to salt, but their roots function as filters, straining most of the salt out of the water they absorb. Other excess salt is carried to the leaves and excreted onto their surfaces.

Because their roots are anchored in mud that is anaerobic, or oxygen-poor, mangrove trees have difficulty obtaining an adequate supply of oxygen. Black mangroves possess special roots that project above the water or that become exposed to the atmosphere at low tide. These customized growths, called respiratory roots or pneumatophores, are dotted with lenticels—small, ventlike openings that take in air and channel it to the parts of the root that lie buried beneath the mud.

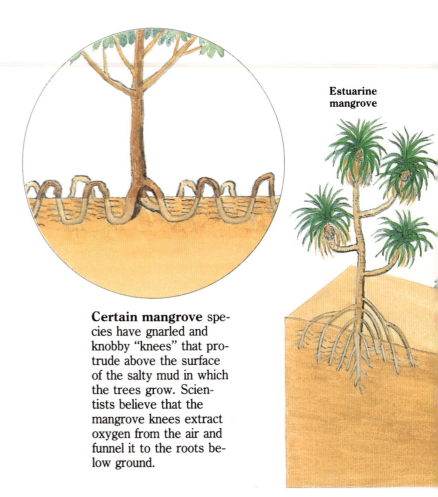

Estuarine mangrove

Certain mangrove species have gnarled and knobby "knees" that protrude above the surface of the salty mud in which the trees grow. Scientists believe that the mangrove knees extract oxygen from the air and funnel it to the roots below ground.

Seeds that are raring to grow

Mangrove seeds get a jump on life by sprouting while they are still attached to the tree. The bean-shaped root may grow a foot long before it drops from the tree. The root then sticks into the mud beneath the tree, or it may be washed to a distant shore.

Fruit →

Young leaves →

Root →

Pores to get the salt out

Salt gland (magnified)

Cell groupings called salt glands help plants like mangrove and cypress trees rid their systems of excess salt. The salt is filtered out by the roots or transported by the sap to the glands and excreted onto the leaf surfaces.

Some mangrove species sprout prop roots *(right)* as the trees get older. The prop roots buttress, or hold up, the growing tree.

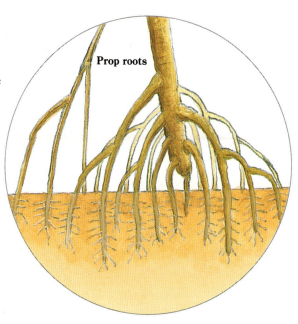

Prop roots

A gallery of mangroves

Mangrove species differ widely in the amount of salt and flooding they can tolerate. As shown below, some live in tidal flats almost always above the water line, while others flourish in deeper water.

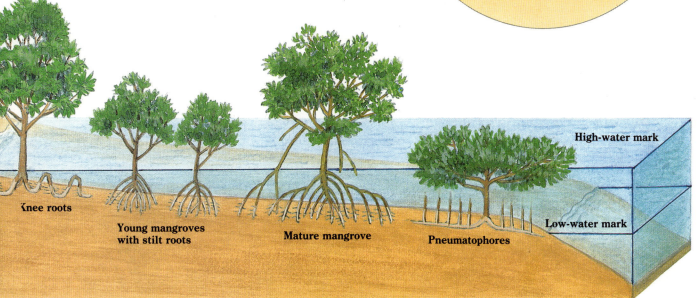

Knee roots

Young mangroves with stilt roots

Mature mangrove

Pneumatophores

High-water mark

Low-water mark

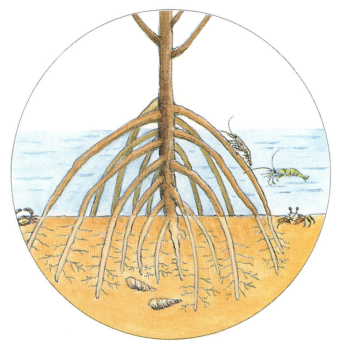

By capturing floating organic debris—including their own leaves, 4 tons of which they shed per acre each year—mangroves help create new land. They also provide habitat for crustaceans.

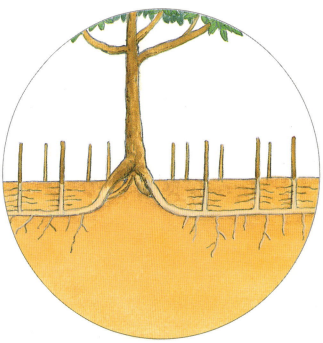

Pneumatophores—literally "bearers of air"—protrude upward from horizontal mangrove roots. These slender extensions serve as the tree's respiratory organs, carrying oxygen to the buried roots.

What Lives on the Forest Floor?

Denizens of the dirt

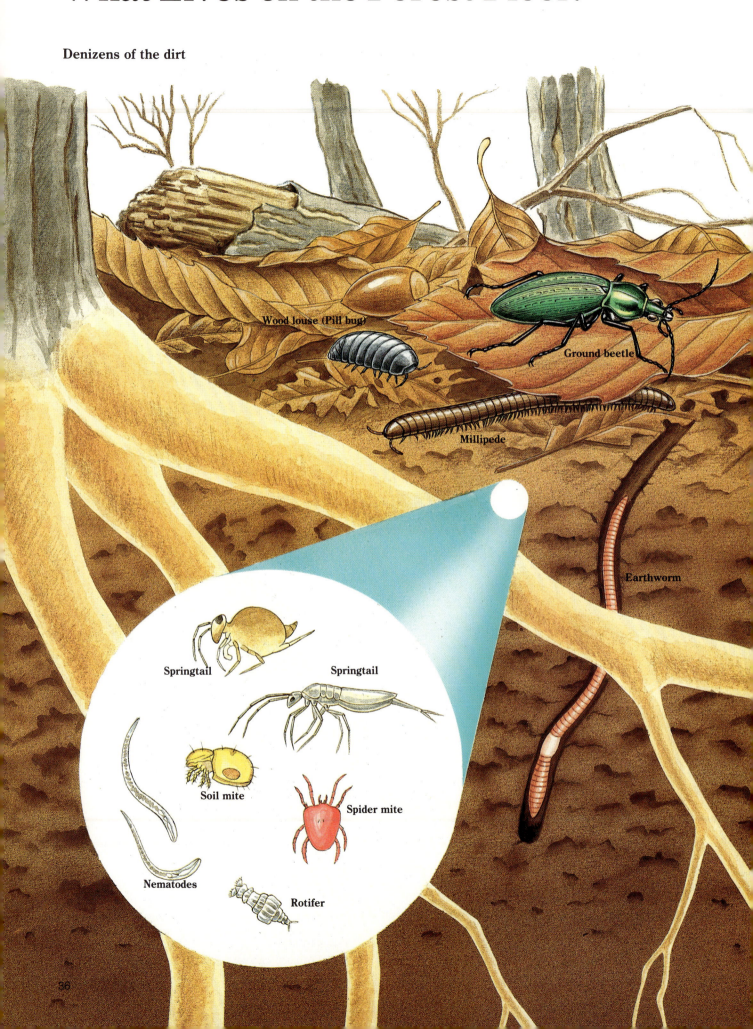

Wood louse (Pill bug)

Ground beetle

Millipede

Earthworm

Springtail

Springtail

Soil mite

Spider mite

Nematodes

Rotifer

The soil in which deciduous trees grow is home to thousands of kinds of living organisms, a handful of which are shown below. Microscopic paramecia and single-celled algae cling to particles of soil. Ground beetles, fireflies, ladybirds, mites, millipedes, ants, wood lice, and pseudoscorpions scurry among the litter, or leafy debris, that carpets the forest floor. Earthworms squirm through the dirt, nourishing themselves and aerating the soil—that is, opening burrows that allow oxygen to penetrate below the surface.

Other creatures, such as moths and cicadas, emerge from the soil as flying insects after spending their larval stages curled in the ground. Such larvae represent a rich source of food for larger animals, notably moles and shrews. Without the activities of all these creatures eating, excreting, reproducing, dying, and decomposing, the earth might be no more than a barren wasteland. From soil literally comes the stuff of life.

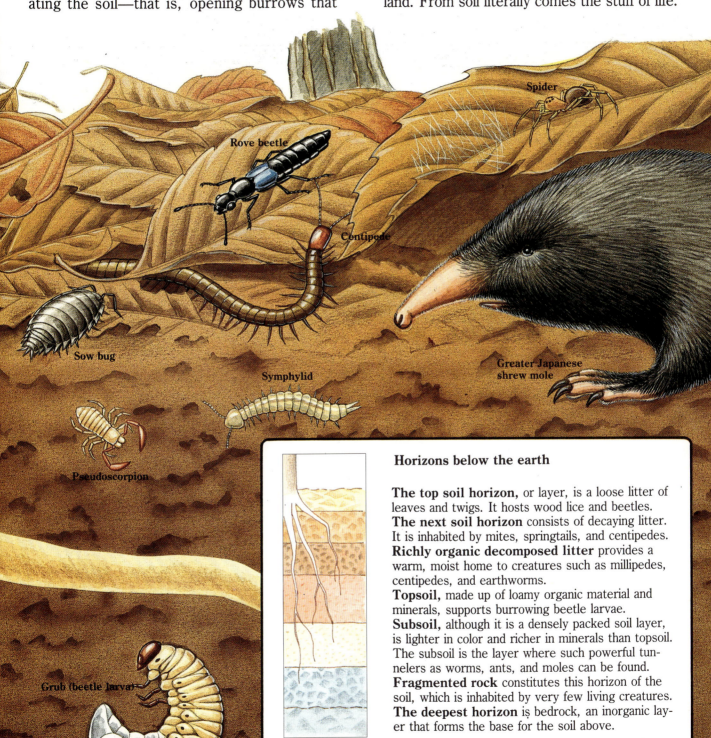

Spider

Rove beetle

Centipede

Sow bug

Symphylid

Greater Japanese shrew mole

Pseudoscorpion

Grub (beetle larva)

Horizons below the earth

The top soil horizon, or layer, is a loose litter of leaves and twigs. It hosts wood lice and beetles.
The next soil horizon consists of decaying litter. It is inhabited by mites, springtails, and centipedes.
Richly organic decomposed litter provides a warm, moist home to creatures such as millipedes, centipedes, and earthworms.
Topsoil, made up of loamy organic material and minerals, supports burrowing beetle larvae.
Subsoil, although it is a densely packed soil layer, is lighter in color and richer in minerals than topsoil. The subsoil is the layer where such powerful tunnelers as worms, ants, and moles can be found.
Fragmented rock constitutes this horizon of the soil, which is inhabited by very few living creatures.
The deepest horizon is bedrock, an inorganic layer that forms the base for the soil above.

Ecologists classify soil according to its horizons, or layers, each of which exhibits its own distinct appearances and qualities. The cutaway view above shows, from top to bottom, leaf litter, decaying litter, decomposed litter or humus, topsoil, subsoil, fragmented rock, and bedrock.

Why Do Bats Live in Caves?

Bats are nocturnal creatures, hunting in the nighttime darkness and sleeping during the day. To be safe from prowling predators as they sleep, many kinds of bats roost in the dark, damp inner recesses of caves. For this reason many people have never seen a bat, even though bats are one of the most numerous groups of mammals with more than 900 species throughout the world. Most bats dwell in the tropics, but they are found in every region except Antarctica and the Arctic.

In climates where the winters are cold, bats are seen even less often in the chilly months. Many species of bats pass the winter, when food is scarce, hibernating in caves. Large numbers of the animals hang upside down from the cave ceiling *(right)*, clinging to it with the claws on their toes. When the weather changes and temperatures climb, the bats may leave the cave for the season or continue to use it as a daytime roost. Some caves have been home to bats for thousands of years.

Bats are the only mammals that can fly. Although some squirrels can glide from tree to tree, they cannot initiate free flight as bats do. Bats' wings, as seen in these flying horseshoe bats, consist of a membrane of skin stretching from their sides to the tips of their finger bones. Horseshoe bats, common in Europe and Africa, have wings well suited for slow flight while hunting near the ground in the forest or bush.

The horseshoe bat is named for the horseshoe-shaped structure on its face. Scientists believe these nose leaves are part of its echolocation navigating system. Unlike many bats, the horseshoe bat emits high-frequency sounds from its nostrils, not its mouth, and uses the echoes returning from nearby objects to help it fly safely in the dark.

For daytime sleep in the summer, horseshoe bats fold their wings beside their bodies *(right)*. But for hibernation *(far right)*, each bat wraps its wings around its body. This snug position helps conserve the moisture that is essential to the bat's survival.

Temperatures in a cave

Scientists have recorded and compared the temperatures at various points inside and just outside a typical cave. As the graphs at right show, caves make ideal roosting and hibernating places for bats, because they maintain a fairly uniform temperature, as opposed to conditions outside. Equally important for bat survival is the constantly high humidity found inside a cave.

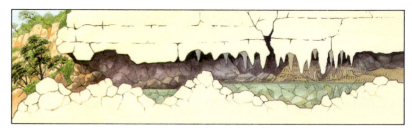

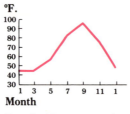

Just inside a cave, the temperature varies.

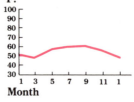

In the middle, there is little change.

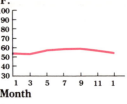

Deep inside, the temperature stays steady.

Bat communication

Pregnant female horseshoe bats move to a special, warmer nursery area of the cave roof. There they give birth—usually to just one offspring—and care for their young. At first, the baby bats go along for the ride when their mothers leave the roost to hunt for food. But as they grow bigger, the young are left clinging to the cave roof in the evenings, when the mothers go out of the cave and feed. Upon returning, each mother bat issues a high-pitched call to its young, and the baby squeals back. By such calling, each baby bat guides its mother to it. When the mother reaches its own young, it nurses the baby at its chest.

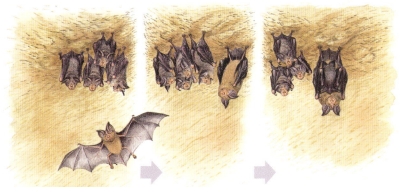

A mother bat leaves its offspring and heads out to feed.

Mutual calling enables mother and baby to find each other.

A nursing youngster clings to its mother to take a meal of milk.

Where Do Cave Dwellers Find Food?

Within the dark reaches of seemingly empty caves, numerous species of plants and animals cling to a meager existence. The barren, sunless caves cannot produce green plants, which are at the base of most food chains above ground. Cave dwellers dependent on fiber or organisms of a complex food chain have to feed outside—bats, for instance, leave at night to hunt. But the permanent cave dwellers must live on the limited food supply inside. Over millennia the caves' food-poor environment has given rise to an intricate network of interdependence, each organism an irreplaceable link in the food chain. Nature also conserves strength by not wasting energy on useless physical features. Many cave creatures have lost their eyes and the pigments that provide protective coloring, neither of which are needed in the dark caves.

Cave food pyramid

On the bottom layer of this simplified aquatic food pyramid are the smallest life forms: the single-celled animals called protozoa, and bacteria and fungi—the parasitic lower plants that lack chlorophyll. These are eaten by small freshwater crustaceans and flatworms of the middle level. In turn, the creatures of the middle level are consumed by the blind crayfish and cavefish at the top of the food pyramid.

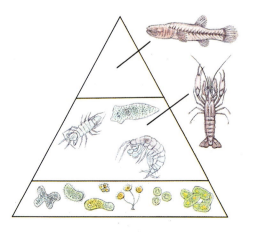

4. Cave crayfish devour copepods, flatworms, and plant debris.

5. Cave shrimp are both prey and predator in cave streams.

6. Copepods and flatworms eat bacteria and protozoa, providing a food source for larger animals.

7. Protozoa and other microscopic cave dwellers consume organic debris and fungi.

2. Dead animals, such as the shrew below, provide nutrients for beetles and springtails. Fungi and bacteria aid decomposition.

3. Organic debris washes into cave streams, providing nutrients for microscopic invertebrates, which in turn are eaten by larger aquatic animals, such as flatworms.

1. Rich in nutrients, bat droppings, called guano, are consumed by beetles and millipedes, and nourish parasitic fungi.

Copepods

Flatworms

How Does the Cave Climate Affect Where Cave Dwellers Live?

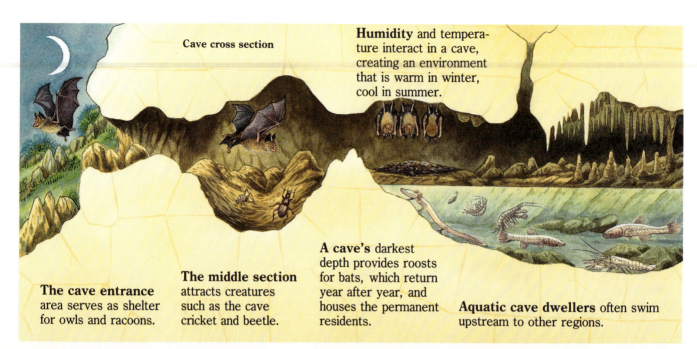

Cave cross section

Humidity and temperature interact in a cave, creating an environment that is warm in winter, cool in summer.

The cave entrance area serves as shelter for owls and racoons.

The middle section attracts creatures such as the cave cricket and beetle.

A cave's darkest depth provides roosts for bats, which return year after year, and houses the permanent residents.

Aquatic cave dwellers often swim upstream to other regions.

Cave shrimp

The blind salamander retains its gills even as an adult.

The blind, colorless crayfish is a typical cave dweller.

The blind cavefish finds its prey with specialized cells on its sides and snout to detect motion.

Life for cave dwellers varies, depending on where they live in a cave. A cave usually has three well-defined regions. The dimly lit area just behind the entrance is called the twilight zone. Here, animals called troglophiles, or cave lovers, seek shelter but leave the cave to find food. Owls and raccoons may be found here. The next region extends from the twilight zone toward the middle of the cave. This region is dark, and the temperature fluctuates slightly. Bats venture here to escape the summer's heat or winter's cold. Some permanent cave dwellers—blind beetles and blind cavefish, for instance—occasionally move into this region. Far inside the cave is the constant temperature zone, where the humidity is high and the temperature remains stable. Cave dwellers prefer this zone, since they need humidity to stay alive. The often sightless creatures depend on sensitive antennae, or feelers, instead of eyes.

The illustration below shows a composite of cave life from different parts of the world.

Life cycle in a food-poor environment

The scavenger beetle of France, like other insects living in caves, goes through a shortened development process. The female lays a large egg, which evolves into a last-stage larva that immediately forms a cocoon and develops into an adult. Surface-dwelling beetles, by contrast, follow a multilarval, multimolt pattern.

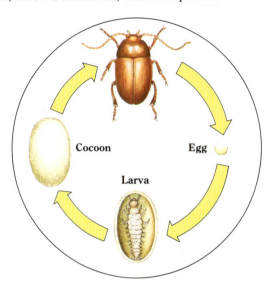

Cocoon

Egg

Larva

Scavenger beetles

Blind millipede

Blind beetle

K. MURAKAMI

Which Plants Grow in Antarctica?

Although more than 95 percent of Antarctica is locked in ice year-round, the remaining 275,000 square miles are home to a few tenacious plants. Only two of these plants—one a grass, the other an herb—produce flowers. The rest are primitive mosses, lichens, algae, and liverworts, which can adapt to harsh habitats like ice fields or survive being frozen for months on end.

During the polar midsummer, when the continent is bathed in sunlight up to 24 hours per day, Antarctica's plant life stores up energy for the coming winter by performing photosynthesis almost around the clock. Then, as light levels and temperatures begin to fall, the plants lapse into dormancy. Buffeted by high winds and beset by extreme cold and drought, Antarctica has one of the planet's most inhospitable climates for higher forms of plant life.

■ **Antarctic survivors**

Leathery *Umbilicaria* and droopy *Usnea* lichens grow in tandem on exposed rock.

▼ **Lush carpets of moss,** like the green *Ceratodon* below, grow on sandy or gravelly soil near water.

Antarctic plant distribution

Antarctica is made up of ice shelves *(green at right),* or sheets of ice projecting into the ocean, and solid land *(white),* most of which is covered year-round by ice and snow. Almost nowhere on the continent does the mean monthly temperature rise above freezing. It therefore seems miraculous that so many plant species—about 800 in all—have acclimated themselves to this harsh habitat. Some have become proficient absorbers of solar radiation. Others have found niches for living in the continent's microclimates: A sheltered rock face, for example, protects plants from the wind, while the edge of an ice sheet offers a steady supply of meltwater. Minute quantities of meltwater can sustain plants such as the slow-growing lichens.

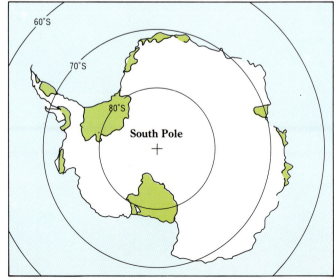

60°S

70°S

80°S

South Pole

+

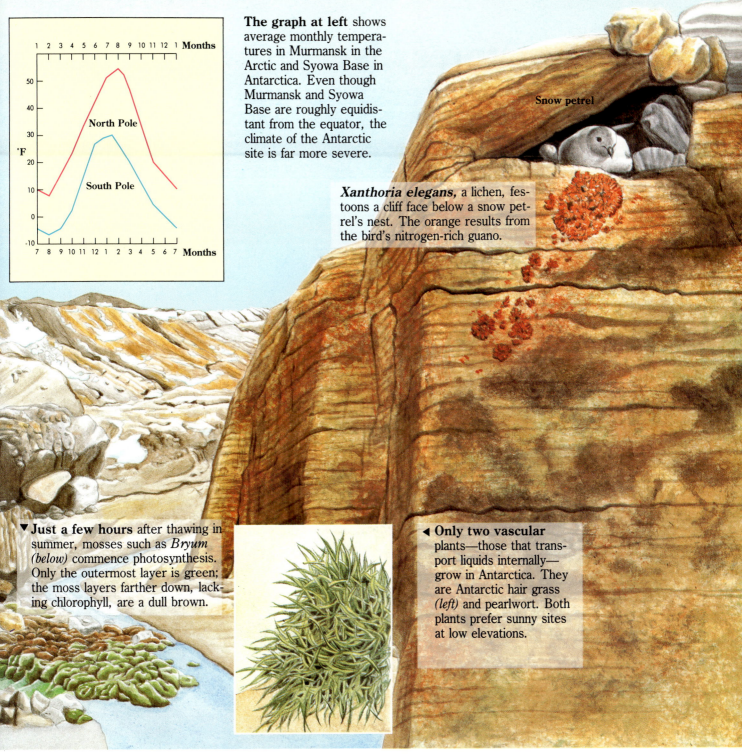

The graph at left shows average monthly temperatures in Murmansk in the Arctic and Syowa Base in Antarctica. Even though Murmansk and Syowa Base are roughly equidistant from the equator, the climate of the Antarctic site is far more severe.

Snow petrel

Xanthoria elegans, a lichen, festoons a cliff face below a snow petrel's nest. The orange results from the bird's nitrogen-rich guano.

▼ **Just a few hours** after thawing in summer, mosses such as *Bryum (below)* commence photosynthesis. Only the outermost layer is green; the moss layers farther down, lacking chlorophyll, are a dull brown.

◀ **Only two vascular** plants—those that transport liquids internally—grow in Antarctica. They are Antarctic hair grass *(left)* and pearlwort. Both plants prefer sunny sites at low elevations.

Moss colony

Undulating beds of Antarctic moss *(above)* are composed of several species. Over the years, they build up into thick mats.

Cross section of moss

This cutaway clump of moss reveals a young green upper layer involved in photosynthesis and an older brown base no longer exposed to sunlight.

45

Which Creatures Live in Antarctic Waters?

The chill waters of Antarctica's oceans support a surprising abundance of plants and animals. Four times more species thrive in these waters than elsewhere. The conditions here have stayed relatively stable for thousands of years, allowing for the development of a diverse marine ecosystem. This wealth of marine life is nourished by phytoplankton, minute floating plants, brought to the water's surface by currents. The phytoplankton support dense populations of tiny shrimplike krill, which in turn sustain numerous larger marine animals like penguins, seals, and whales.

At depths below 100 feet, temperatures are cold but stable and food is scarce. As a result, the animals function at low metabolic rates, growing more slowly but larger than similar species in warmer waters. Here sponges can live for several centuries. Sea anemones are anchored to the seabed, and octopus, starfish, isopods, and sea spiders feed nearby. Another bottom dweller, the ice fish, like several other polar fish, has adapted to the cold by producing up to eight different kinds of antifreeze molecules to keep the body fluids liquid at below freezing temperatures.

Swimming through the pack ice, seals are well insulated with dense fur and a thick layer of blubber. Of the five seal species in Antarctica, the crabeater seal is the most abundant, with a population estimated at 20 million to 40 million. Despite its name, the crabeater feeds mainly on krill. The predatory leopard seal frequents the waters off penguin rookeries but also feeds on krill and young crabeater seals. Penguins are the most abundant Antarctic seabirds. The Adélie penguins shown here are one of five penguin species found in these waters. During breeding season, the Adélies congregate in colonies of more than 100,000.

Phytoplankton

Shown many times enlarged, phytoplankton serve as the main plant food source in Antarctic waters. Microscopic in size, they bloom, or multiply, rapidly during the Antarctic spring and summer between October and February.

Antarctic petrel

Leopard seal

Seasnail

Sea anemones

Sea urchins

■ Antarctica's abundant marine life

Antarctic terns

Adélie penguins

Phytoplankton

Krill

Crabeater seals

Antarctic herring

Octopus

Brittle starfish

Giant isopod

Sponge

Sea spider

Ice fish

Starfish

Sea squirt

47

How Do Animals Survive in Polar Areas?

The Arctic and Antarctic experience bitter cold winters, often lasting up to 10 months a year. Temperatures dip down to −60° F. in Antarctica and down to a slightly less frigid −30° F. in the Arctic. Yet each region supports rich ecosystems. More than two dozen species of mammals live in the Arctic. Antarctica has no native land mammals and only a few invertebrates and plants, while the oceans abound with marine life. Taken together, the polar regions are home to the largest populations of seabirds in the world; to large numbers of aquatic plankton, whales, walruses, seals, polar bears, and herds of caribou; as well as to muskoxen and lemmings. They manage to survive the harsh climate through adaptation and specific behavior. Some species migrate to a more temperate environment during the winter, others hibernate, while the rest have developed special features to protect them against the cold.

Clustering for warmth

Muskoxen have double-layered coats. The outer coat of long, coarse hair is insulation against snow and rain; the inner layer is so dense, it holds in the body heat. Their broad hoofs keep the animals from sinking into snow and help them dig for food.

Lemmings' hideout

Typical of many small mammals in the Arctic tundra, lemmings burrow beneath the snow to make a nest, padded with dry vegetation, and build a network of interconnected tunnels. Here they store food for the winter and wait out the coldest months.

● **Fur for ice and water**

The Arctic polar bear's white coat serves both as camouflage while it stalks its prey and as thick insulation. Beneath its fur, the bear has a thick layer of fat that also helps to keep the animal warm.

Cozily tucked inside, polar bear cubs snuggle up to their mother in a snowbank den. In the fall, females dig dens in which they will give birth and keep the young until spring.

Hardy animals of the Antarctic

Antarctica's birds and mammals endure the cold through adaptive insulation such as dual layers of feathers or fur and blubber. Some species also use special behavioral strategies to survive.

Male emperor penguins gather into a tight group for warmth. In strong winds, those on the outside change places with others from the inside after a short watch.

An underwater survivor, the Weddell seal spends the winter months submerged. The seal gnaws holes through the ice so that it can come up for air.

What Do Reindeer Eat?

Reindeer, called caribou in North America, roam in herds throughout the northernmost regions of North America, Greenland, Europe, and Asia, making yearly migrations to their summer and winter ranges. In spring and summertime, reindeer graze on young grass shoots, fungi, and other vegetation, traveling up to thousands of miles in search of fresh greenery. During this period, females shed their antlers, give birth, and nurse their calves, while males begin regrowing antlers that they shed after mating in the fall. When snow covers the ground, reindeer must settle for less nourishing food in the form of lichen. Reindeer scrape snow away with their broad hoofs to reach the lichen beneath, then nip off pieces with their small teeth. Lichen grow slowly and could easily become overgrazed, but reindeer move on after a time, leaving the plants to regrow.

A winter's meal for reindeer

Old-man's-beard lichen grows from tree branches in northern forests, a food source for antlerless male reindeer when snow is on the ground.

Summer's bounty

Reindeer find more food sources in summer, including willow and birch leaves, berries, grasses, and mushrooms, which provide energy and help fuel growth.

Willow

Blueberry

Sedge

Mushroom

Male antlers in summer

Pawing at snow with its hoof, a reindeer exposes the lichen underneath. Lichen are made up of two types of plant: 90 percent fungi and 10 percent algae, which produces starch through photosynthesis.

Females, which have smaller antlers than mature males, will share feeding holes with their calves, but occasionally, like males, may fight each other for a chance to get to an already exposed patch of lichen.

Female and calf

Lichen

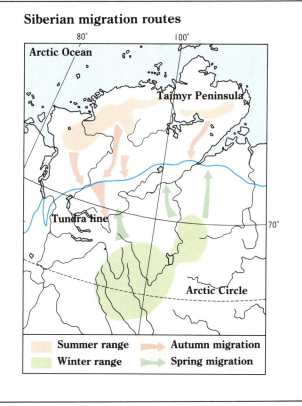

Siberian migration routes

Arctic Ocean

Taimyr Peninsula

Tundra line

Arctic Circle

| | Summer range | | Autumn migration |
| | Winter range | | Spring migration |

Annual migration routes for reindeer of north-central Siberia are shown above. Summer finds the herds moving northward.

How Did Alpine Butterflies Evolve?

High peaks around the world are home to a distinctive group of butterflies called Alpines. This is surprising because butterflies have poor mechanisms for regulating body temperature, so they are not often found in cold latitudes. Scientists speculate that Alpine butterflies probably evolved during the last ice age. As huge sheets of ice, sometimes more than a mile thick, moved down from the Arctic regions, the ranges of many plant and animal species were pushed southward. Some butterflies were stranded on nunataks, those few bits of land—sort of mountaintop islands—that remained above the icy flood. Having colonized these regions, they would have remained when the ice retreated.

In another possible scenario, the ancestors of Alpine butterflies would have adapted to the lower temperatures of the ice age. Then, as the earth warmed, those species would have sought cooler regions and colonized mountain valleys.

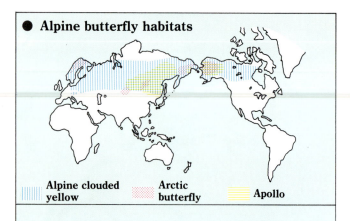

● **Alpine butterfly habitats**

Alpine clouded yellow   Arctic butterfly   Apollo

The map above shows distribution patterns for three common Alpine butterflies, the Alpine clouded yellow, or *Colias palaeno;* the Arctic, or *Oeneis norna;* and the Apollo, or *Parnassius palaeno.* These are found in the high mountains of the Himalayas, the Alps, Scandinavia, and the Rockies, and also in Alaska and Siberia.

■ **Alpine butterflies**

Parnassius eversmanni

Fritillary butterfly

Rhododendron

Bleeding hearts

Parnassius eversmanni

Heather

Arctic butterfly

One evolutionary scenario

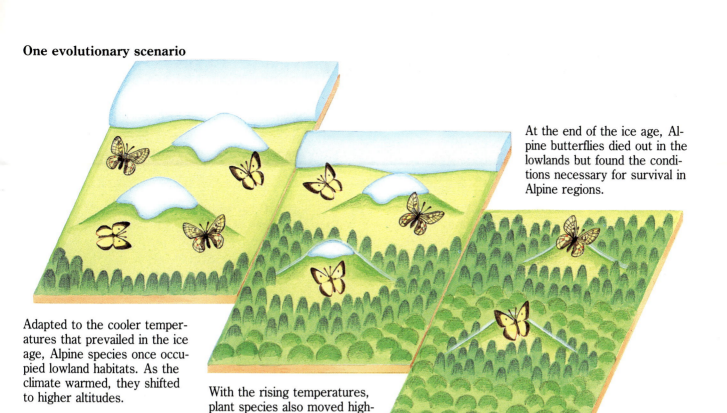

At the end of the ice age, Alpine butterflies died out in the lowlands but found the conditions necessary for survival in Alpine regions.

Adapted to the cooler temperatures that prevailed in the ice age, Alpine species once occupied lowland habitats. As the climate warmed, they shifted to higher altitudes.

With the rising temperatures, plant species also moved higher up on mountain slopes, creating the ecosystems that were just right for the butterfly populations.

Alpine clouded yellow

Creeping pine

Clossiana freija asahidakeana

Blueberry bushes

Life cycle of *Parnassius eversmanni*

Third year

First year

Second year

Lowland butterflies complete their life cycles in a year, but Alpine butterflies, forced to adapt to summers in northern latitudes, which are too short to allow them to mature to adulthood, take three years to go through a cycle. The eggs survive the winter in the first year and hatch in spring as caterpillars. The caterpillars feed during the short summer then endure the next winter as pupae in cocoons and emerge the following spring as adults.

Which Trees Grow at Timberline?

On high mountains, true trees do not grow above a certain altitude. The timberline defines this zone. Just below the timberline grow only the hardiest trees, that is, the coniferous cedars, junipers, pines, and firs. Above the timberline—which varies in elevation depending on latitude, wind conditions, and exposure to sunlight—only stunted shrubs, low-growing grasses, flowering herbs, mosses, and lichen can survive. Not even these plants are able to grow on the bare rocks on mountain peaks. Scientists assume that the extremely short growing season between the last and the first frost, in the Alpine tundra lasting between 50 and 90 days, is the key factor in the lack of trees above the timberline. The longer a winter's cold persists, the less time plants have to make food through photosynthesis and thus the less energy they have to grow. Only smaller plants can cope with such conditions. Tundra species have also developed extensive root systems to help hold water in gravelly mountain soils and tiny leaves to retain moisture.

Snow-laden pine

Rhododendron

Wind-distorted branches

▲ **Conifers such as pine, fir, cedar,** hemlock, juniper, and larch grow just below the timberline, which may vary in altitude from mountain to mountain depending on wind and specific climate. Unlike Arctic tundra, Alpine tundra—found above the timberline—receives a lot of sun in the winter. Extreme cold and the shortness of the summer growing season are therefore the main factors that prevent trees from surviving in the higher mountain regions. Heavy snows also take a toll, snapping and warping branches.

◄Creeping pine

▼**In the Alpine tundra** just above the timberline can be found low-lying shrubs like the creeping pine, shown below and, weighted by snow, at right in the closeup. Other tundra plants include some heathers, sedges, grasses, and dwarf birches and willows, none more than a few feet tall. Subjected to a lack of water, high winds, and cold, these plants grow extremely slowly.

▲**A transition zone** between the coniferous forest and the Alpine tundra, sometimes called the subalpine boundary, supports sparse stands of trees mixed with shrubs, like the wind-swept rhododendrons shown at left. These highly stressed trees often twist into gnarled shapes.

▼**Below the timberline**

On lower mountain slopes, trees grow in dense stands, like the Japanese larches above.

▼**Above the timberline**

At higher elevations, the same trees may grow entirely differently, as illustrated by the creeping larches.

55

What Are Alpine Meadows?

■ **An Alpine meadow in bloom**

Cinquefoil

Crane's-bill

Gentian

Alpine plants compared to warm-climate plants

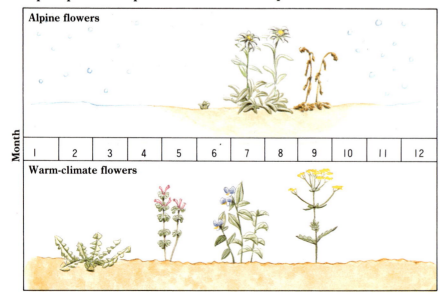

Alpine flowers

Month

| 1 | 2 | 3 | 4 | 5 | 6 | 7 | 8 | 9 | 10 | 11 | 12 |

Warm-climate flowers

In mid-May, snows begin to melt, and Alpine plants sprout rapidly. During the brief summer, plants form leaves, bloom, and produce seeds. September frosts kill the upper portions, but the roots, with stored energy, survive the winter.

In warmer zones, spring arrives earlier, and plants begin to grow. Both annual and perennial plants have time to produce seeds. Come autumn, annuals begin to die. Their seeds will sprout in the next year. In the mildest climates, perennials may continue growing year-round.

Although Alpine regions suffer harsh conditions, including extreme winter cold, high-speed winds, and intense sun, they support a wide variety of flowering plants. In spring, in treeless areas, mountain peaks are covered with a brilliant array of color. In these so-called Alpine meadows, hundreds of perennial species flourish, most with small leaves, short stems, and bright blooms. Red and blue tints are most common as a result of the presence of pigments called anthocyanins. These are beneficial to cold-weather plants, allowing them to produce heat from sunlight in spring and to resist freezes in winter. Most Alpine flowers form the following year's buds by fall. That way, as soon as spring arrives, the blooms are ready to open, leaving as many days of warm weather as possible for seeds to mature and new buds to form.

Parnassius eversmanni

Bleeding heart

Anemone

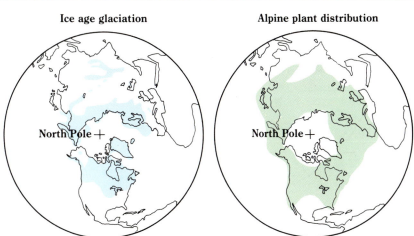

▲ In tundra regions, the growing season is often too short for annuals to bloom and set seed. Perennials store energy in roots from which they regenerate in spring, thereby speeding up their growing cycle.

Alpine plant distribution

The map at right shows the extent of glaciation (blue) during the last ice age; the map at far right the distribution of Alpine plants (green). Modern species trace their origins to the colder climate that prevailed during the ice age. When the glaciers retreated, these species colonized higher elevations where conditions suited their genetic makeup.

Ice age glaciation

North Pole +

Alpine plant distribution

North Pole +

What Do Moose Eat?

Moose, the largest members of the deer family—sometimes reaching 7 feet at the shoulder and weighing up to 1,800 pounds—range through forested regions of the northern latitudes. Bull moose have large, flattened antlers and mate with females, called cows, in autumn. Moose feed on both terrestrial and aquatic vegetation. In winter and summer, they mainly eat leaves and tender twigs of hazel shrubs, dogwood, mountain maple, willow, and other trees, stripping branches with their pendulous lips. This food, which is passed through the moose's four rumens, or stomachs, for digestion, provides high levels of energy. In summer, moose also seek out grasses and aquatic plants, such as pondweed, water lilies, and horsetail, from lakes and streams. Wading in, they submerge their heads and grab mouthfuls of the bulky vegetation, lifting their heads as they chew to scan their surroundings for wolves or bears, which are their main enemies.

Moose can run swiftly and, because of their large bodies and long legs, can clear obstacles as much as 3 feet high without jumping when pursued. Wolves or bears, on the other hand, become fatigued quickly in a chase on uneven terrain. Once a moose is full grown, even a pack of wolves cannot kill it, unless it is old or weak.

A moose's diet

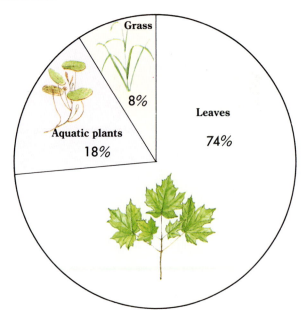

Grass 8%

Aquatic plants 18%

Leaves 74%

Moose feed only in the cooler hours of the day and may eat the equivalent of 20,000 leaves or 1,000 aquatic plants a day. Leaves provide bulk, while grasses furnish nutrients for energy and for storage as fat for the cold winter months. Aquatic plants supply salts that are converted to amino acids, important for the growth of connective tissue, hair, and horn.

Moose feeding on aquatic plants

Why Are Tropical Birds So Colorful?

In temperate climates, a bird's coloring often acts as camouflage, which helps to protect it from predators. In the dense, deeply shaded jungles of the world, even birds with brilliantly colored feathers are difficult to spot among the lush vegetation. Rain forest birds rely on size and speed of flight to escape their enemies. In mating rituals, males with the brightest plumage attract the most females. Yellow, red, and black pigments in the feathers cause the gaudy colors, and interlocking feather barbs that function like venetian blinds create the birds' iridescent appearance. Tropical birds have also developed features to take advantage of jungle conditions, including oversize bills for eating or amplifying their calls; long, narrow beaks for reaching nectar; short wings for navigating in close quarters; and pincerlike feet for climbing.

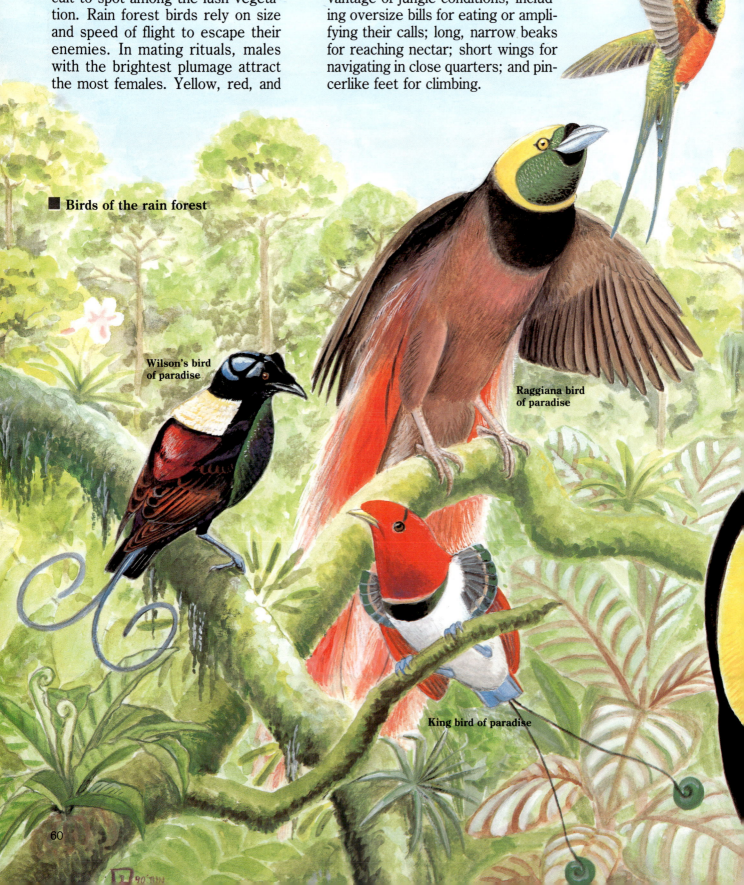

Fiery topaz

■ **Birds of the rain forest**

Wilson's bird of paradise

Raggiana bird of paradise

King bird of paradise

Broad-billed
hummingbird

Scarlet tanager
in summer plumage

Blue-crowned
motmot

Crimson-breasted
barbet

Great jacamar

Lettered toucan

Collared trogon

Scarlet macaw

Keel-billed toucan

What Are the Habits of the Sloth?

The name of this tropical creature comes from the Old English word for slow. Indeed, the two-toed and three-toed sloths are among the slowest-moving animals in the world. They spend much of their lives asleep, clutching tree branches with their curved, hooklike claws. During their few waking hours, the tailless animals clamber hand over hand upside down along branches or hang in spots where they can easily reach leaves to munch on with their peglike teeth. In this unhurried way, sloths expend little energy. Their coarse brown hair, in which green algae grows at times, lets them blend in with their surroundings. Sloths rarely descend to the ground, where their ungainly walk makes them prey to jaguars and other predators.

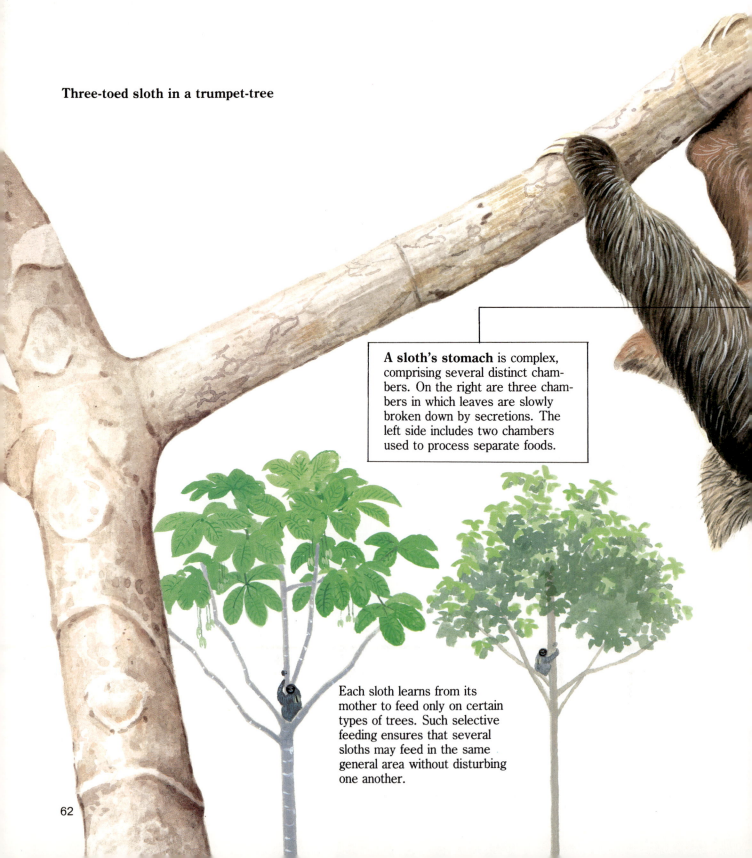

Three-toed sloth in a trumpet-tree

A sloth's stomach is complex, comprising several distinct chambers. On the right are three chambers in which leaves are slowly broken down by secretions. The left side includes two chambers used to process separate foods.

Each sloth learns from its mother to feed only on certain types of trees. Such selective feeding ensures that several sloths may feed in the same general area without disturbing one another.

Central
stomach

Right
stomach

Esophagus

Esophageal opening

Left stomach

Stomach
chamber

Pylorus

To intestines

Stomach chamber

Care of the young

For the first month, newborn sloths cling to their mothers to nurse. Sloths give birth to a single offspring in summer.

After a month, the young sloth begins to feed on leaves. Within six to nine months, the mother cedes part of its feeding range to its offspring.

Which Trees Grow in Rain Forests?

Tropical rain forests, which lie along the earth's equatorial zone, are among the most diverse habitats in the world. These warm, drippy environments harbor thousands of species of plants, many not yet identified by science. Tropical plants that are useful to people include hardwoods like teak, mahogany, rosewood, and ebony; food trees like those that yield coffee, cocoa, and nutmeg; and scores of medicinal plants. Most tropical trees are evergreens, and many flower year-round, some blooming several times in succession. Although shallow-rooted, the tallest trees reach 200 feet, with the majority standing about 130 feet high. Few branches interrupt the trunks on their reach toward sunlight. The unusual conditions in the rain forest have forced many strange adaptations in the trees' development.

1. Some species of rain forest trees send out arching roots above the ground. Taking hold, these wide-flying stilt roots help prop up tall trees like lines staking a tent.

2. Boardlike buttress roots are a type of prop root that not only ensure stability but also aid in the intake of water and oxygen.

3. Many rain forest trees are cauliflorous; that is, they put forth leaves, flowers, and fruit directly from their trunks or branches.

4. The stranglers begin life as epiphytes, which take nutrients from air and rainwater through aerial roots. Sending roots to the ground, they may soon girdle and kill mighty trees.

5. Cauliflorous fruit, like the breadfruit, grow from blossoms pollinated by birds, insects, and bats in the dark understory of the forest.

6. Lianas, which are rooted in the ground, creep up nearby trees to reach the sun at the top of the rain forest. Young lianas may look like shrubs that only later become vines.

3

5

6

4

How Are Rain Forests Structured?

Rain forests can be thought of as having four layers. Beginning at the top, the emergent layer is made up of one or two extremely tall trees per acre. Mostly small-leaved with flat crowns, these trees are exposed to strong winds. Next comes the canopy, reaching between 65 and 130 feet. Here, crowns of leathery leaves overlap in a thick layer, capturing sunlight but allowing some of the rain to trickle down to lower levels. This is where most animals, including monkeys, tree shrews, lemurs, birds, and many insects spend their lives. Beneath the liana-draped canopy, at 15 to 65 feet high, stands the dimly lit understory of young trees, palms, shrubs, and epiphytes. Finally, there is the moist forest floor with its leaf litter, mosses, and ferns—home to mammals, insects, and reptiles.

The world's tropical rain forests

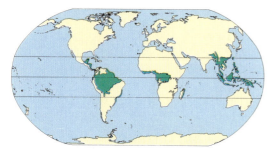

The largest remaining tracts of rain forest occur in Central America, the Amazon Basin, western equatorial Africa, Madagascar, parts of southeastern Asia, and New Guinea.

A tufted canopy of green

In the photo above, the lush rain forest canopy is unbroken except for the meandering track of a river. Here and there, emergent trees protrude above the rounded crowns of the canopy trees. Thousands of species may be found in a single swath of tropical forest.

The layers of a rain forest

The leaves of the canopy absorb 99 percent of the sunlight, shielding the lower parts of the forest, where the average annual temperature and humidity are constant.

The tallest trees exist in a rarefied zone above the canopy. Asian rain forests boast giant emergent trees, with some more than 200 feet tall.

Emergent layer

Canopy

Understory

Forest floor

Why Do Many Rain Forest Plants Have Air Roots?

In the understory of a rain forest, light is dappled or dim. Although moisture is plentiful, it is mostly available in the form of water vapor rather than liquid water. At some levels, both temperature and humidity may rise and fall rapidly each day. To take advantage of these conditions, epiphytes—plants with air roots—evolved. Epiphytes have no need to root in soil; they grow on sheltered branches instead, or affix themselves to tree trunks, and absorb nutrients and water from the air. Some epiphytes have tangled roots that act like sponges or traps to catch decaying leaf matter that provides nutrition. Others have swollen pseudobulbs, or root reservoirs. About 9,000 species of orchid are epiphytic, as are many bromeliads.

▶ **Insects** pollinate showy *Dendrobium* orchids.

◀**The pseudobulbs** of Asian *Coelogyne nitida* store water.

▲ *Vanda* orchids from southern Asia also grow in dry areas.

◀ **Beneath** the thick-leaved *Drymoglossum,* ants live in harmony with the plant.

▲ **Staghorn ferns** can survive severe water loss. They may wilt but revive when rewetted.

◄ **One stem** of Asia's *Phalaenopsis,* the moth or- chid, may hold 100 blooms.

A miniclimate

The rain forest's canopy is like an umbrella, block- ing sunlight and rain. Un- derneath the canopy, an intricate balance of hu- midity and gases is main- tained. The leaves may total 10 times the sur- face area covered by the forest itself. Through transpiration, they re- lease moisture, which bathes the forest floor. They also absorb sunlight and carbon dioxide, fuel- ing photosynthesis. In return, the leaves re- lease oxygen.

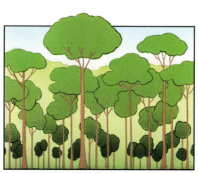

Sunshine   Humidity   Rainfall   Temperature   Wind flow

Maximum   Minimum

Emergent trees and the canopy inter- cept more than 90 percent of the sun- light, allowing less than 1 percent to reach the floor. Humidity, on the oth- er hand, is trapped by the leaves and rises to almost 100 percent near the ground by night and drops to 50 per- cent by day. The leaves also catch rain so that little actually reaches the ground. And the sun's heat bounces off the canopy so that the lower lay- ers remain cooler. Wind whips the emergent trees above but scarcely causes a breeze on the forest floor.

Why Do Some Monkeys Live in Trees?

Gibbons

■ **The forest acrobats**

Woolly monkeys

Orangutans

Squirrel monkeys

Arboreal, or tree-dwelling, monkeys and apes of the tropical rain forests spend nearly all their time in treetops, foraging for food, sleeping, grooming, and even giving birth. In the dense canopy, they swing from branch to branch, feeding on insects, fresh green leaves, tender shoots, and fruit. On the forest floor, they would be prey to jaguars and other hunters, whereas in the treetops their only major predators are eagles and other birds of prey.

The monkeys on these pages are not drawn to scale and represent Old World species of Asia and Africa and New World monkeys of Central and South America.

Howler monkeys

Black-and-white colobuses

Spider monkeys

Lion tamarins

Bodies built for trees

Gibbon

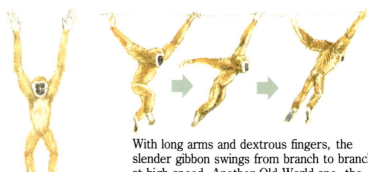

With long arms and dextrous fingers, the slender gibbon swings from branch to branch at high speed. Another Old World ape, the much larger orangutan, also travels by moving hand over hand through the forest.

Squirrel monkey

Squirrel monkeys of the New World are squirrel-size, with long legs and whiplike tails to aid them in balancing, much like the even smaller lion tamarins. Old World colobus monkeys fling themselves similarly along.

Spider monkey

Tiny spider monkeys and larger woolly and howler monkeys of the New World use prehensile tails with rough ridges of hair like a fifth limb for support.

Why Are Rain Forests Disappearing?

The ruination of a rain forest

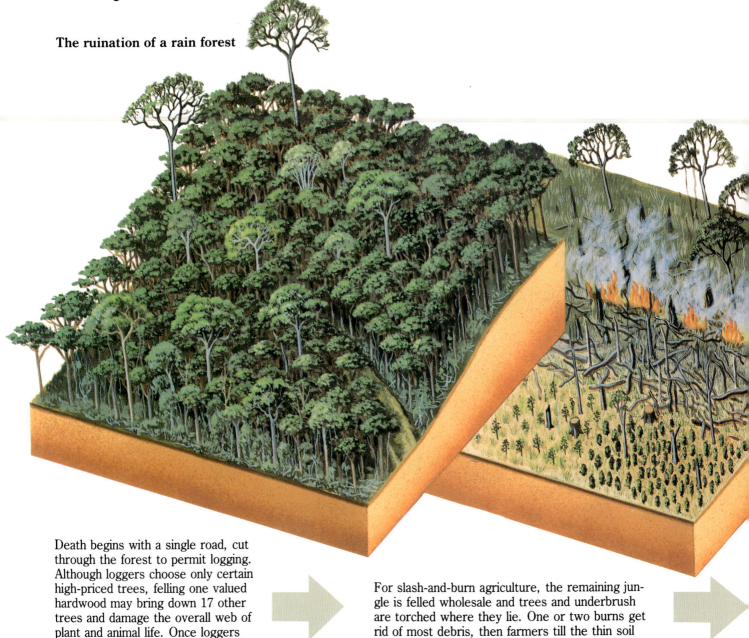

Death begins with a single road, cut through the forest to permit logging. Although loggers choose only certain high-priced trees, felling one valued hardwood may bring down 17 other trees and damage the overall web of plant and animal life. Once loggers are finished, farmers move in.

For slash-and-burn agriculture, the remaining jungle is felled wholesale and trees and underbrush are torched where they lie. One or two burns get rid of most debris, then farmers till the thin soil and sow crops like cassava, sugarcane, and corn.

Death by logging

Only select trees in the rain forest are sought for lumber. But any logging effort unavoidably ruins large areas of forest. In Indonesia, for example, commercial logging in a given area eats up 30 percent of the forest. Of this amount, roads and outbuildings account for one-third, actual tree felling for another third, and secondary tracks made by tractors hauling out logs for the rest. Such scars may take more than 50 years to heal, but the forest may never again be as productive either ecologically or economically.

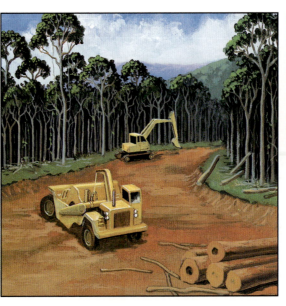

The mountainside above, in Thung Salaeng Luang National Park in Thailand, was once forested, but slash-and-burn farming has denuded it in just one decade.

Forests once spread over more than a fourth of the world's land; today they cover only one-fifth. In Central and South America, Africa, and Asia over the past 20 years, billions of acres of rain forest have been destroyed by slash-and-burn agriculture, cattle grazing, logging, and mining. The reasons are many for this pillaging of one of the earth's most vital resources; construction of dam, road, and resettlement projects in the rain forests has hastened the rate of loss. Population explosions in the Third World have driven peasants into the forests to cut down the trees and carve out new farms and cattle ranches. Ironically, the soil that supports the most diverse ecosystems yields little fertility when stripped of its vegetation. New crops cease to produce after a few years. Then, wind and water erode the land, turning it into a desert. Local rainfall decreases and temperatures soar, adding to the warming of the planet by the greenhouse effect.

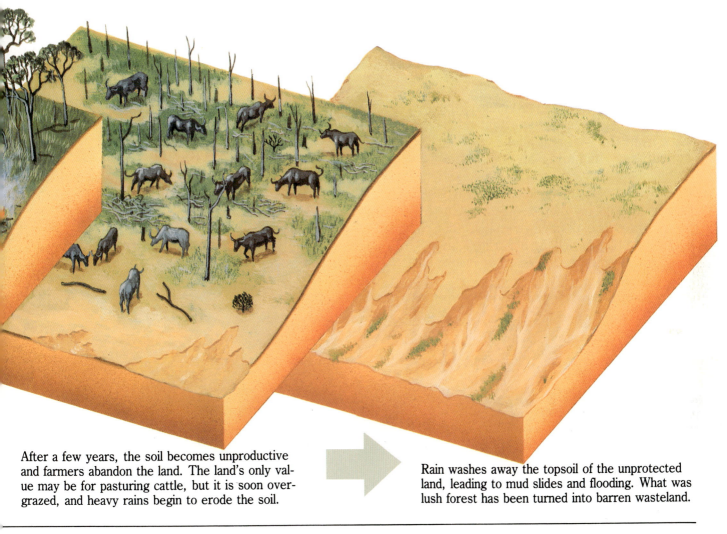

After a few years, the soil becomes unproductive and farmers abandon the land. The land's only value may be for pasturing cattle, but it is soon overgrazed, and heavy rains begin to erode the soil.

Rain washes away the topsoil of the unprotected land, leading to mud slides and flooding. What was lush forest has been turned into barren wasteland.

Apocalypse along the Amazon

Brazil's 1,980-mile Trans-Amazon highway, begun in 1970, made all but the remotest parts of the rain forest accessible to humans. Relocated by the government, hundreds of thousands of Indians and peasants were forced to settle in the newly opened territories, where they eked out a meager existence from slash-and-burn farming. Wealthy investors, including Americans and Japanese, also bought up huge tracts for grazing. Today, many people in the region's 400 settlements are impoverished, and huge stretches of the rich, biologically diverse rain forest have vanished forever.

The 1973 satellite photo above shows the Amazon Basin as the highway was being built. Small tribes of Indians lived in harmony with the forest, farming patches of land in a way that did little long-term harm.

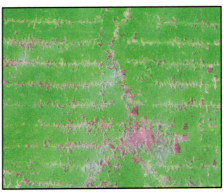

Red patches in this 1986 photo indicate clearings, branching off from the highway, where large swaths of the forest had been cut. Untold numbers of plants and animals had perished or become extinct.

What's Killing Temperate Forests?

In the 1970s, scientists discovered that forests throughout Europe and North America were dying, especially those with large stands of the conifers spruce and fir. Atmospheric testing revealed that pollution from power plants and automobiles was to blame. Chemicals called sulfur dioxides and nitrogen oxides were rising into the atmosphere, combining with other molecules, and falling to the ground as precipitation 10 times more acidic than normal.

Sometimes as strong as vinegar, the acid rain was leaching important nutrients from the soil and allowing other substances, including aluminum and manganese, to harm roots. Weakened trees were being attacked by insects and disease, losing their needles. In Europe, whole forests have given in to this man-made blight, making acid rain a serious environmental problem.

■ **An acid-etched landscape**

Death in the Black Forest

In the photo at right, taken in 1970, fir and pine in southern Germany had already begun to show subtle signs of decline. Ten years later, as shown at far right, the area was almost barren. By 1985, more than 55 percent of Black Forest conifers had been affected by acid rain. Now, deciduous trees such as beeches are failing. A German scientist has called this an ecological "catastrophe of previously unimaginable dimensions."

Black Forest, 1970

Same area, 1980

A menace creeps over Europe

Modern, super-tall smokestacks have spread acid rain far and wide. The diagrams below track acid rain's spread from 1956 to 1985, giving figures for pH by region. A pH of less than 7 shows acidity; the strength of the acid increases tenfold with every one-digit decrease in pH.

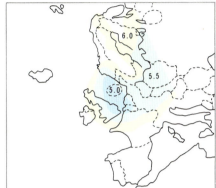

Acid rain in Europe, 1956

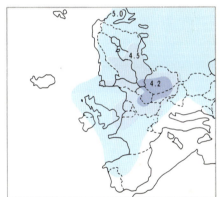

By 1985, the damaging rainfall was more acidic and more widespread across northern Europe.

Reversing the damage?

German scientists have tried to stop the effects of acid rain by sprinkling the ground with a powdery substance called lime. In the experiment shown at right, the alkaline lime neutralized the acid soil and helped grasses and saplings grow. This strategy might help bring back forests, but only if they are protected from acid rain from now on.

In a beech forest, acid rain killed grasses essential to the trees' health.

Three tons of lime, spread on the forest floor, restored the undergrowth.

What Are Grasslands?

In the Central Asian steppes, daily temperature changes can exceed annual ones, as shown below. Annual rainfall totals 18 to 20 inches.

Four types of grassland

Steppes contain grasses, a few low-growing trees, and wildflowers.

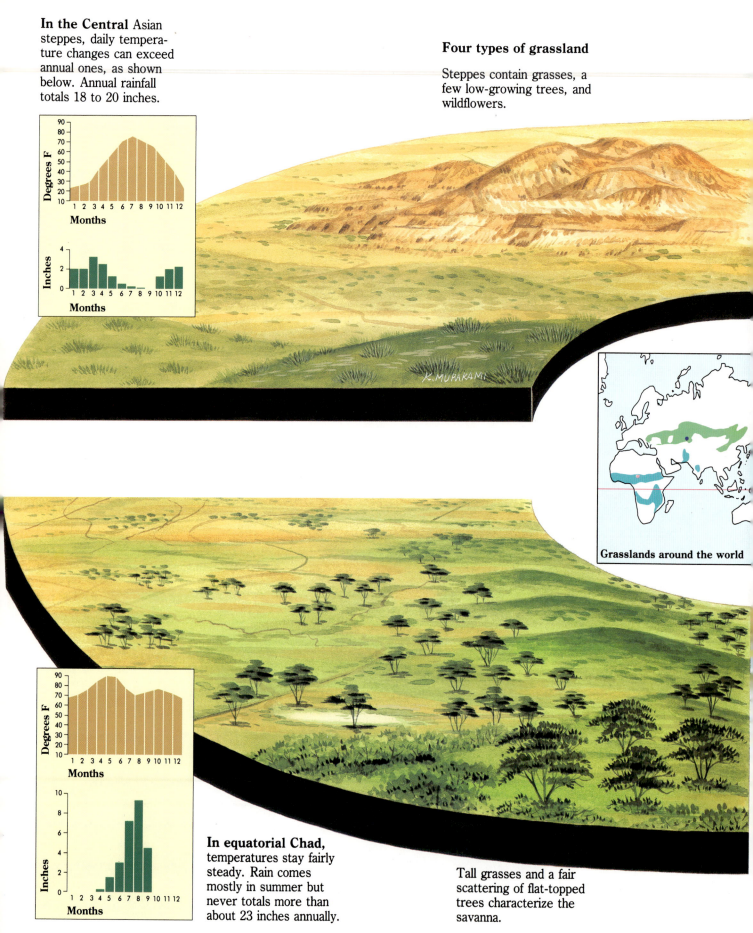

Grasslands around the world

In equatorial Chad, temperatures stay fairly steady. Rain comes mostly in summer but never totals more than about 23 inches annually.

Tall grasses and a fair scattering of flat-topped trees characterize the savanna.

African savannas, Eurasian steppes, North American prairies, and South American pampas all are types of grassland. These great, almost treeless expanses of grasses generally appear where the swing between summer and winter temperatures is wide and annual rainfall is sparse enough to discourage forests but ample enough to prevent the area from turning into a desert.

Perennial grasses, which are highly evolved wind-pollinated plants that return year after year, thrive under such circumstances. Natural grasslands support a wide range of insects, birds, and mammals. Farmed grasslands, such as the cultivated grainfields in America's Midwest, can be highly productive in their own right. Some grasslands require regular fires or they will eventually revert to forest. Bogs, marshes, and Alpine meadows are also considered grasslands, even though they are created and maintained under different conditions.

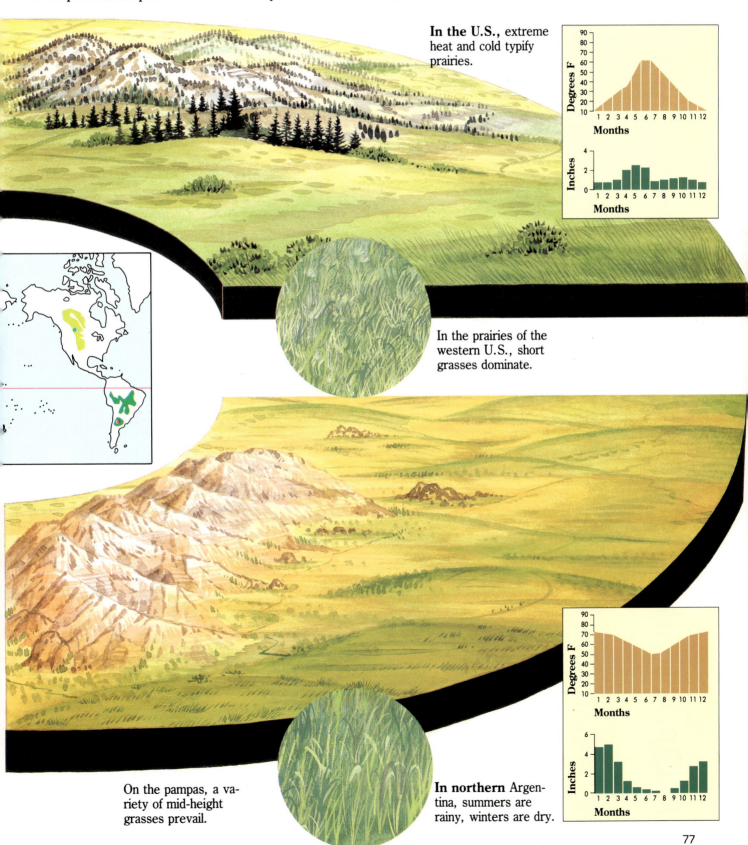

In the U.S., extreme heat and cold typify prairies.

Degrees F

Months

Inches

Months

In the prairies of the western U.S., short grasses dominate.

On the pampas, a variety of mid-height grasses prevail.

In northern Argentina, summers are rainy, winters are dry.

Degrees F

Months

Inches

Months

How Do Grasses Regrow after Fires?

From time to time, fires engulf grasslands. In these natural blazes, plant life appears to be destroyed. Generally, however, portions of some plants remain beneath the soil, and from these buried stems and roots, the plants can grow again. Also, seeds singed in the flames often sprout new seedlings. In fact, some species even need exposure to intense heat before their seeds will germinate. Fires tend to weed out certain plants and foster the survival of others, especially grasses, colonies of which will suppress the growth of bushes, shrubs, and trees.

Plants destroyed by fire

Fire-tolerant plants

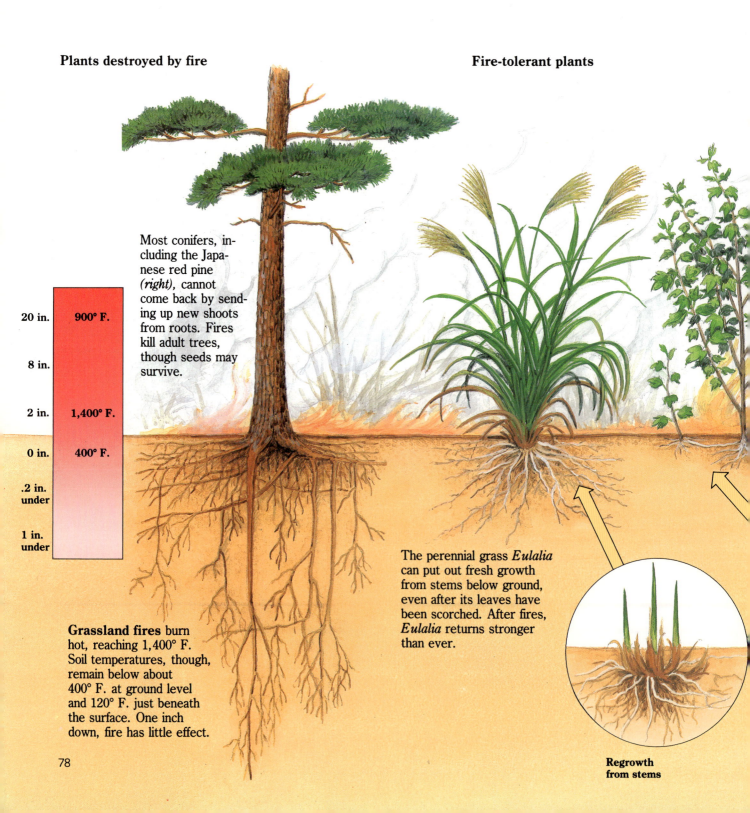

20 in.	900° F.
8 in.	
2 in.	1,400° F.
0 in.	400° F.
.2 in. under	
1 in. under	

Most conifers, including the Japanese red pine *(right)*, cannot come back by sending up new shoots from roots. Fires kill adult trees, though seeds may survive.

Grassland fires burn hot, reaching 1,400° F. Soil temperatures, though, remain below about 400° F. at ground level and 120° F. just beneath the surface. One inch down, fire has little effect.

The perennial grass *Eulalia* can put out fresh growth from stems below ground, even after its leaves have been scorched. After fires, *Eulalia* returns stronger than ever.

Regrowth from stems

Seeds that like heat

Fires often encourage dormant seeds to sprout. Heat is vital to the reproduction of many plants, including the clover *Lespedeza*, the vine *Pueraria*, and the shrub *Rhus javanica*, whose seeds will not respond merely to water. A well-known example in the United States is the lodgepole pine: The cones of this species can only open after fire, releasing their seeds into an environment in which there is little competition from other plants. In South Africa, one species of fireball lily sprouts from its bulb only after a fire, producing brilliant red flowers.

***Lespedeza* quickly carpets** a mountain after a fire.

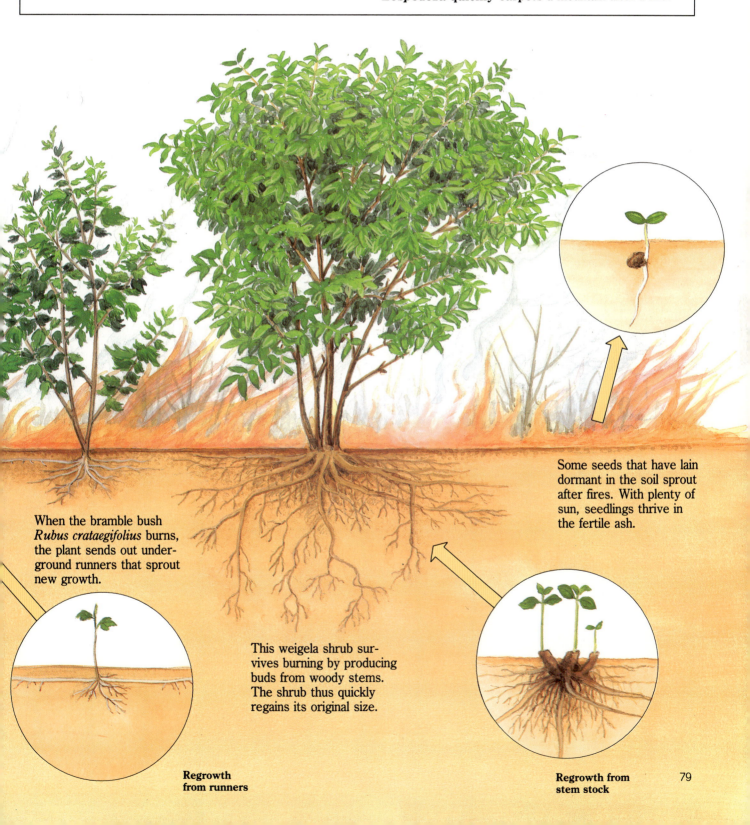

When the bramble bush *Rubus crataegifolius* burns, the plant sends out underground runners that sprout new growth.

This weigela shrub survives burning by producing buds from woody stems. The shrub thus quickly regains its original size.

Some seeds that have lain dormant in the soil sprout after fires. With plenty of sun, seedlings thrive in the fertile ash.

Regrowth from runners

Regrowth from stem stock

79

Why Do Zebras Live in Herds?

Evolutionary biologists are fascinated by the instinct of creatures to flock together. Butterflies swarm, fish swim in schools, and elephants form herds, each for different reasons. Insects such as mayflies, for example, gather to mate, with males flying in thick clouds over water to attract females. Bats fly in groups for feeding, easing the search for food.

Large grazing animals like zebras herd for protection from predators. In any gathering of zebras, a lead stallion and several watchful others keep a lookout for lions or other enemies approaching across the savanna. Upon sighting an enemy, the guard stallion trumpets an alarm, and the herd charges away with males at the back protecting mares and foals. In the rush, predators may become distracted or deflected and may give up the chase.

A grazing herd of zebras

In a herd, many pairs of eyes are devoted to watching for attackers, thus raising the chances of survival for all members of the group.

A charging lioness becomes confused as its striped prey dash away in every direction.

Why Do So Many Large Mammals Live on the African Grasslands?

The savannas of Africa provide not only a great amount of food for herbivorous (plant-eating) animals but also a great variety of plants. As a result, large creatures have thrived there, each species evolving to feed on different kinds of plants. Impala, for instance, browse on woodland bushes, while African buffalo favor swampier parts of the grasslands. If several species eat the same plant, they do so at different stages of the plant's growth. Thomson's gazelles, for exam-

Predators in the grass

When a predator looms, female elephants circle around their young, staving off attackers.

Thomson's gazelles *(above left)* respond to a cheetah by stotting, or leaping stiff-legged into the air. The tiny dik-dik *(above right)* is well camouflaged and hides among the bushes, unseen by a passing hyena.

Herbivores

Ostrich

Carnivores

Lion

Black rhinoceros

Cape hunting dog

Spotted hyena

ple, graze on young grass shoots, but zebras prefer the tough outer stems of mature grass.

The huge herds of herbivores, in turn, provide ample meat for carnivorous beasts, from the great hunter, the lion, to the scavenging hyena.

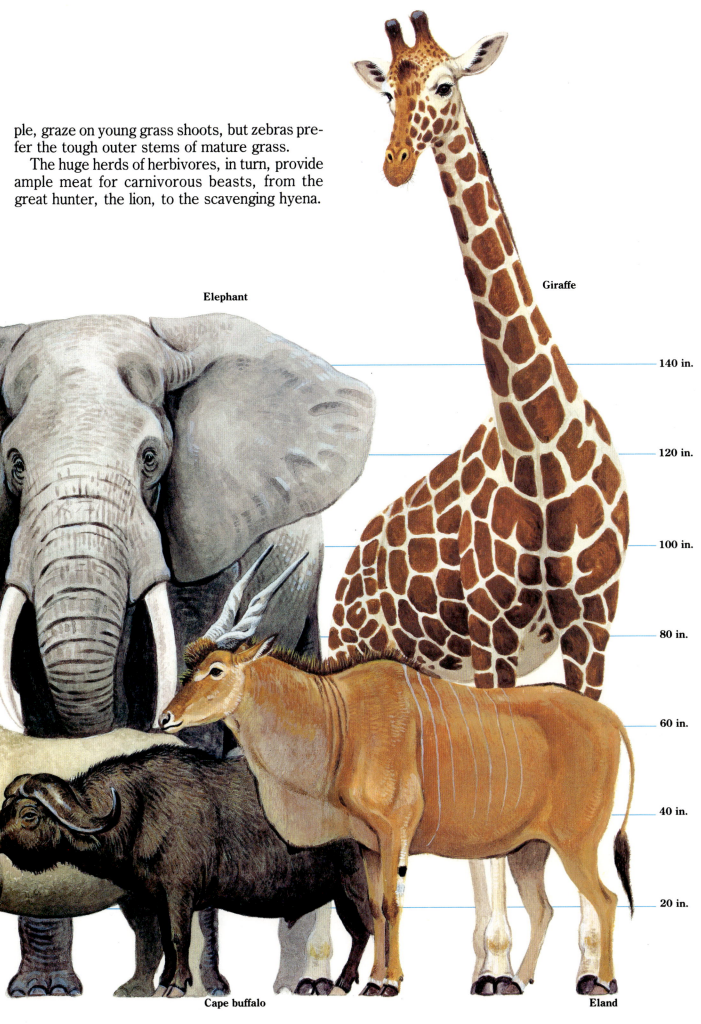

Giraffe

Elephant

Cape buffalo

Eland

140 in.

120 in.

100 in.

80 in.

60 in.

40 in.

20 in.

What Kinds of Food Do African Herbivores Eat?

Zoologists, or scientists who study the animal kingdom, have found that the big plant eaters of Africa generally eat according to their weight. Tiny dik-diks and pygmy antelopes, weighing about 11 pounds fully grown, nibble on the leaves of shrubs. Bushbucks, between 70 and 170 pounds, survive on shrubs and grass. And topis and gnus, weighing more than 220 pounds, eat only grass.

Rhinos and elephants can eat a wider variety of plants. The white rhinoceros lives along forested riverbanks where tasty shrubs are plentiful. Elephants require as much as 550 to 750 pounds of food a day. They are not at all picky about what they eat, grabbing leaves, tree bark, branches, fruit, aquatic plants—virtually any plant material that is in their path.

▼Black rhino

◄Gerenuk

▲Gnu

▲Warthogs

■ African herbivores
at a water hole

▼ Elephants

◀ Giraffes

◀ Waterbuck

▲ White rhino

▲ Zebras

Topi ▶

▲ Eland

85

Why Do Deserts Bloom All at Once?

In the Sonoran desert of the southwestern U.S., cacti such as upward-reaching saguaro, tentacled ocotillo, and beavertail compete with ground-hugging plants for scarce water. For most of the year, conditions are too dry to permit the plants to bloom.

Middle Eastern deserts are the most difficult ecosystems on earth for plant life. A few hardy species store their own water; others react instantly to rainstorms, run through their life cycles in a few days, and then go dormant for months or years. Only around oases can vegetation thrive.

In each of the earth's distinct hot deserts, a time comes during the year when all the plants seem to bloom at once. This spectacular occurrence is tied directly to the harsh reality of deserts, which typically receive an annual rainfall of about 8 inches and experience temperatures that average 80° F. in the summer. When rains come to the desert, often in late spring, plants have a mere three months or so in which to bud, grow, flower, produce fruit, and go to seed. Evolution has dictated that only those plants that produce seeds during the short wet season survive.

When the rains come to the southwestern desert, typically in both winter and summer, the austere landscape bursts with color. For just a few weeks, flowers appear on the tips of cacti, and the desert floor is covered with a carpet of purple lupines.

In the rocky deserts of central Australia, infrequent rains flush out flowers from clumps of Spinifex grass. Purple parakilya flowers come to life, and the bright red blooms of Sturt's desert pea awake from long dormancy.

How Do Animals Stand Desert Heat?

Desert animals have evolved impressive methods for dealing with the intense heat and dryness of their environment. Many smaller creatures hunker down in dens or shady areas during the hot part of the day, emerging to hunt or browse at dusk or later at night. This is because their small bodies absorb heat rapidly; with too much sunlight, their temperatures would skyrocket.

Other creatures have adapted by conserving water, which helps cool their systems. In reptiles, thick skin and slow rates of metabolism keep water loss low. Other animals, such as the camel, are able to tolerate extreme dehydration. They can survive losing 30 percent of their normal water content—whereas human beings will die upon losing more than 13 percent.

Changes in body heat

Dehydrated

Well-watered

°F
106
104
102
100
98
96
94
92

12 0 12 0 12 0 12

Time

A dromedary, a member of the camel family, withstands wider shifts in its internal temperature than most animals; its body heat can range from 93° F. to 105° F. *(chart above).* Because its temperature drops so far during the night, the camel stays cooler for a longer time when daylight comes. In this way, it keeps moisture loss through perspiration quite low.

Dromedary

During the day, the sand lizard burrows into the relatively cool subsurface of the desert with its webbed feet. Hoods over its eyes and nostrils keep sand grains out.

Near the desert surface, day and night temperatures vary widely. Digging downward, animals enter a zone where temperatures are steadier, swinging about 18° F. 6 inches below surface but only 2° F. at 10 inches.

By day, certain scorpions retreat to their burrows to stay cool, emerging to hunt after dusk.

In the American Southwest, the tiny elf owl nests in saguaro cactus or desert trees by day, soaring forth to hunt insects and their larvae under cover of darkness.

The compact fennec fox of North Africa waits for evening under a shady overhang. Its 6-inch ears help to cool the animal by providing a large surface area through which body heat disperses.

The North African jerboa spends the day sleeping in its burrow. This creature drinks no water, processing what little it needs from seeds it collects.

Can Animals Survive without Water?

Life without water

The whip-tailed kangaroo rat can be found in North American deserts, where it lives in deep, underground burrows. Able to leap like a kangaroo and using its tail for balance, this creature becomes active at night, when the desert cools off. It drinks virtually no water, gaining what it needs from seeds.

Water use of the kangaroo rat

Given a month's supply of food in the form of 3½ ounces of dried barley seed, the kangaroo rat will extract 2 ounces of liquid from it. At the same time, it will lose 2⅕ ounces of water through respiration or excretion. The missing ⅕ ounce of water, which the animal needs to keep its metabolism balanced, comes from breathing the moist air in its nest.

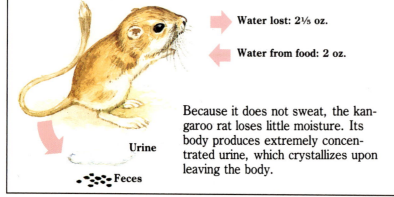

Water lost: 2⅕ oz.

Water from food: 2 oz.

Urine

Feces

Because it does not sweat, the kangaroo rat loses little moisture. Its body produces extremely concentrated urine, which crystallizes upon leaving the body.

The kangaroo rat's burrow *(above)*, 18 inches below ground, provides a home with steady temperatures and humidity throughout the changeable desert day. Although the outside temperature ranges from 50° F. to 140° F., the inside of the den remains at a constant 73° F. Humidity outside may vary from 20 to 60 percent, but interior humidity stays at 60 percent at all hours.

No animal can live without water; however, some can get by without drinking it directly. Some animals extract water from leaves, seeds, grasses, or other substances. Others absorb water through their skins from the atmosphere, while still others convert sugars from food into water, filling their need for fluids.

Since water is vital to life, the metabolism of animals must control how much they lose (through respiration, perspiration, and excretion) compared to how much they take in. Desert creatures have developed the ability to survive the constant scarcity of water, whereas temperate-climate creatures, which lack such abilities, may be devastated when prolonged droughts strike their homes.

Addax are large North African antelopes, the adults weighing up to 295 pounds. Like other desert dwellers, including several other varieties of antelope, they extract water from vegetation. Grazing at dusk, the addax can go for weeks without drinking.

Land tortoises *(right)* carry water with them in a reservoir under their shells.

This desert lizard *(below)* finds liquids and nourishment in insects it catches.

Water reservoir under shell

Reptiles are well suited to desert climates. They get most of their water through their food and lose hardly any moisture through their skin. Like the kangaroo rat, they are able to excrete wastes in highly concentrated form requiring little liquid.

How Do Desert Creatures React to Rain?

Toad and shrimp life cycles

During dry months, adult spadefoot toads remain dormant in holes a foot underground, which they dig with their shovel-like hind limbs. In the cracked mud of a dried-out puddle, brine shrimp eggs sit in a suspended state, neither alive nor dead.

Sensing rain, the spadefoot toads churn to the surface. Males begin to chorus loudly in the brimming puddle to attract mates. Meanwhile, the soaked brine shrimp eggs transform into larvae and begin to swim.

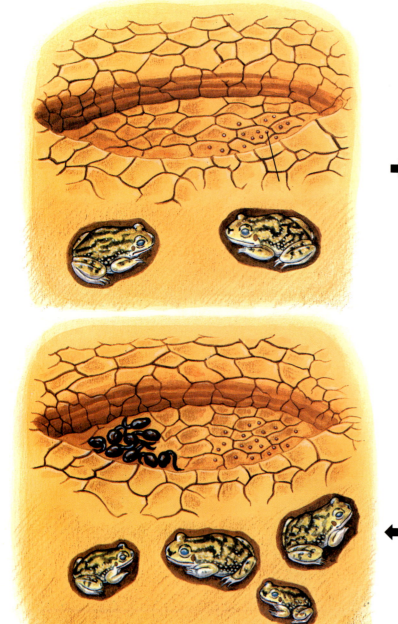

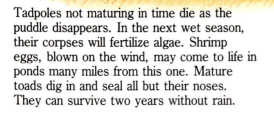

Tadpoles not maturing in time die as the puddle disappears. In the next wet season, their corpses will fertilize algae. Shrimp eggs, blown on the wind, may come to life in ponds many miles from this one. Mature toads dig in and seal all but their noses. They can survive two years without rain.

The puddle evaporates in the heat. If it does not dwindle too quickly, tadpoles have time to become toadlets and hop out. The brine shrimp, however, die as the water wanes.

Some desert species have adapted so that they carry on their business through wet and dry times alike. Others depend entirely on seasonal rainfall to trigger their reproductive cycles. Even an amphibious creature like the frog has relatives in dry climates—some 50 desert species are known worldwide—and these frogs may lie buried in the sand for months, waiting for rain before they can mate. Almost as soon as the first drops of rain hit the ground, these animals shift into high gear, mating, laying eggs, and hatching their rapidly maturing young in a matter of days. The spadefoot toad *(below)* responds to rain, as does the brine shrimp, a crustacean whose seedlike eggs may be blown about in desert winds for as long as 50 years before hatching.

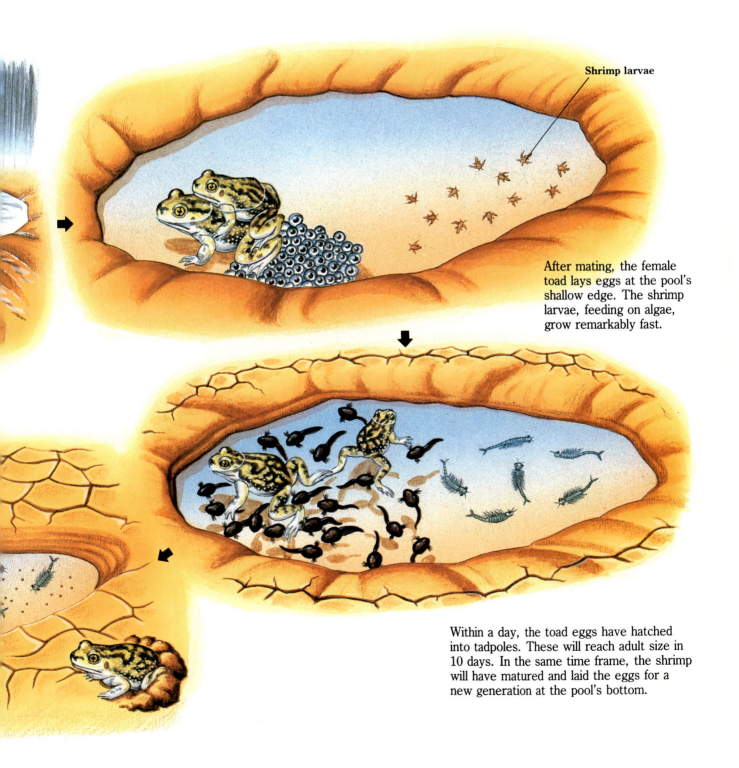

Shrimp larvae

After mating, the female toad lays eggs at the pool's shallow edge. The shrimp larvae, feeding on algae, grow remarkably fast.

Within a day, the toad eggs have hatched into tadpoles. These will reach adult size in 10 days. In the same time frame, the shrimp will have matured and laid the eggs for a new generation at the pool's bottom.

How Do Desert Plants Survive on So Little Water?

In the desert, perennial plants—those that live for more than one growing season—have evolved three basic approaches for dealing with their harsh habitat. They may do one or more of these things: take up rainwater rapidly through masses of tiny roots; store water in their tissues; or live with a low rate of transpiration, the process in which plants lose water through their stomata, or pores.

The acacia shrub relies on the first method,

sending an extensive network of fine roots deep into the desert soil to capture moisture year-round. Cacti and other succulents absorb water in the wet season and store it in their tissues. The saguaro cactus, for instance, holds up to a ton of surplus liquid. Cacti also have minimal transpiration rates. Some species close their stomata during the day to retain moisture. For the same reason, other plants have spines or thick, waxy coverings on their leaves.

Tillandsia cyanea

T. cyanea flower

Water drops

The arched, folded leaves of the *Tillandsia cyanea (left)* act as collectors for dew or mist. After condensing, the water trickles to the base of the stem, where it is absorbed. This method is so efficient that the plant needs almost no root system.

Storing water

During the rainy season, succulents store water in cells in their leaves and stalks. Then they ration it out over the dry season.

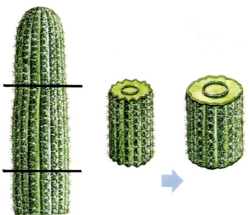

The barrel cactus *(above)* actually swells as it packs water into the cells of its fleshy outer layer. Then, as it uses the excess, it shrinks and its accordion folds show up more clearly.

Some desert species, including the cereus *(above)*, have bulbous roots that serve as reservoirs. In droughts, they drop their leaves completely but can regrow from their roots alone.

Mesquite roots *(right)* descend 20 to 60 feet. This enables the plant to tap deeply buried water even during the driest times.

The branches of the cholla cactus end in spines, not leaves. Many desert plants have evolved such features so as not to lose moisture through evaporation.

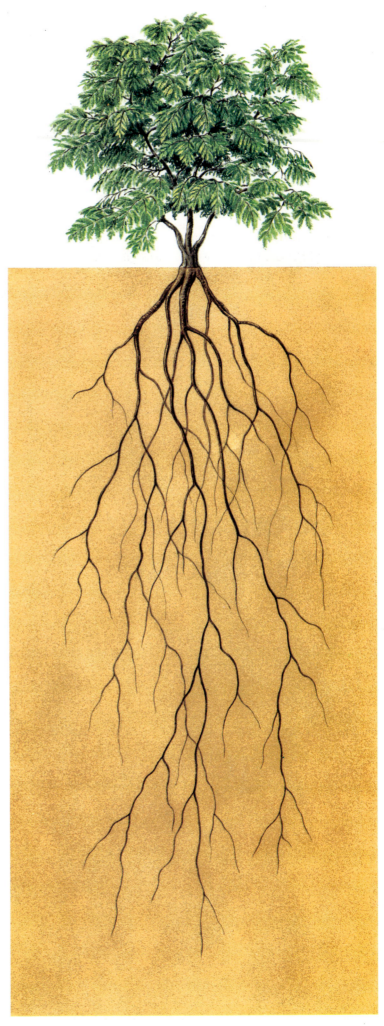

Thick, waxy leaves

The thick, waxy outer coating, or cuticle, of the spiny agave *(above)* and yucca plants prevents water from evaporating from their spongy inner tissues under the heat of the desert sun.

3
Continuing the Species

Behind most animal behavior lies the driving force of reproduction. In their own way, plants also struggle to survive and to produce seeds. Every species strives to create a healthy new generation, sometimes cooperating with other species and other times competing.

Some plants make chemicals that poison nearby competitors, and fast-growing forest trees leave their slow-growing neighbors behind to die in the shade. Some sea animals, competing for ideal rocks to attach to, are so crowded that they grow in strange shapes. Many animals pro-

duce young as quickly as they can—sometimes causing overpopulation. Others bear few young or refrain from reproducing when conditions are crowded or take years to reach maturity.

Each of these behaviors helps to keep a genetic line going. During eons of evolution, successful organisms have overcome many obstacles, while less successful ones have become extinct. But now many plants and animals face a new challenge: Because of the fast-growing human population, constantly encroaching development, and the poisonous by-products of civilization, thou-

sands of species may disappear in the next few decades. Human beings also live in and depend on the web that connects all living things. As people learn more about the web, they may succeed in saving some endangered species and, in doing so, will help the human race too.

Many kinds of animals, including those pictured above, help plants reproduce by carrying seeds and pollen. The rare animals pictured below are members of some of the endangered species that are facing extinction.

How Do Plants Protect Themselves?

Plants cannot chase intruders away, but many do produce chemical defenses. Some of these substances can retard or even prevent the growth of nearby plants. Chemical deterrence may start early, even at the seed stage. Seeds of wheat and other crops, for example, release their own weed-control chemicals that prevent the seeds of other species from sprouting. And once a plant has begun to grow, its leaves and roots may give off substances that block the growth of other species. Sagebrush in the southwestern United States, for example, so poisons the ground that no other plants will grow nearby until there is a fire that destroys the sagebrush and neutralizes the chemical at the same time.

Curiously, a weed-killer produced by a plant can later limit the growth of the same plant that secreted it. When the substance builds up, even the seeds of the plant that created the chemical fail to sprout well in nearby soil.

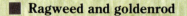

■ **Ragweed and goldenrod**

▲ **Taking over.** Ragweed *(above)* and goldenrod *(below)* overrun fields, keeping other plants from growing nearby.

● **Spreading tall goldenrod**

Plants may seem to secrete poisons for their own benefit, but the example of a weed called tall goldenrod suggests that these chemicals are only by-products of growth. A chemical called poly-acetylene oozes from the roots of tall goldenrod and remains in the soil, hampering the growth of other plants. But after several decades, not even goldenrod seeds can grow in the tainted soil, and the plant slowly disappears.

How do plant defenses work?

Some plants produce chemicals that give the leaves and stems a strong taste or smell—as in the chilies, spices, and herbs used in cooking. All of these substances discourage or even kill insects and other animals that might eat the plant. When rain soaks similar substances into the ground, the chemicals may prevent the growth of other plants that would compete for valuable water, soil nutrients, or sunlight.

In addition, some plant roots secrete chemicals, keeping nearby plants from growing. Such chemical defenses may lead to the patchiness seen in some plant communities. Thistles keep oats from growing, radishes slow the growth of spinach, and walnut trees have been known since the first century AD to kill apple trees.

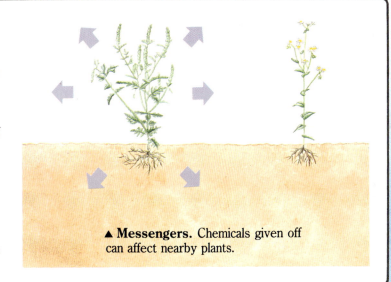

▲ **Messengers.** Chemicals given off can affect nearby plants.

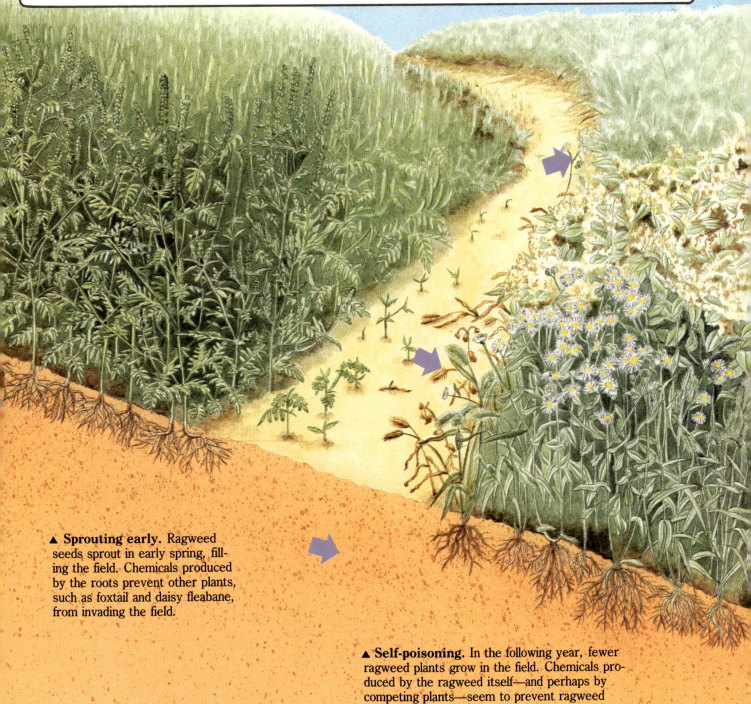

▲ **Sprouting early.** Ragweed seeds sprout in early spring, filling the field. Chemicals produced by the roots prevent other plants, such as foxtail and daisy fleabane, from invading the field.

▲ **Self-poisoning.** In the following year, fewer ragweed plants grow in the field. Chemicals produced by the ragweed itself—and perhaps by competing plants—seem to prevent ragweed seeds from sprouting.

Why Do Some Barnacles Grow Tall?

▼ **Room to grow.** The young barnacles add on to the edges of their cone-shaped shells, increasing the size of the cones. If there is enough room, some barnacle species may grow up to 2 inches in diameter and less than 2 inches high.

Young, crowded colony

Uncrowded colony

The structure of a barnacle

Despite their limestone shells, barnacles belong in the same group as crabs—animals with jointed legs and a tough outer skeleton. With their thin legs, barnacles strain microscopic food from the seawater, reaching out through their two-part lids with sweeping motions. Within the shell, each barnacle has all of the organs needed to carry on life—including both male and female sex organs. Despite this, barnacle eggs are usually fertilized by a neighbor nearby.

Barnacle seen from above

Lid

Shell plate

Internal structure of a barnacle

Feeding legs

Shell

Cross section of muscle

Shell plate

Stomach

Intestine

Anus

Crowding causes some barnacles to grow as long and thin as a large crayon. Barnacles begin life in the ocean as tiny swimming larvae. As they mature, they choose a suitable surface—rocks, dock pilings, boat bottoms, and even whale flippers will do—and settle down to build a hard shell. Then they never have to move again.

The barnacle larvae prefer to settle on surfaces where there are others of their kind. As a barnacle grows, it continually adds to the edge of its pale shell, usually forming the familiar low volcano shape *(below, center)*. But on ideal surfaces, some barnacle species crowd together so tightly that as they grow, their shells bump together. Since they cannot spread out, their shells become tall and tubular.

▼ **Growing up in a crowd.** On the most favorable surfaces, nearly 1,400 barnacle larvae pack a square foot. Under these conditions, barnacles reach only about ½ inch in diameter before their shells touch, but they continue to grow outward, sometimes stretching up to 4 inches in height.

How barnacles attach

After hatching inside its parent, a larva swims to the bottom in search of a home. In a few days, two glue-producing glands grow near its antennae, enabling it to attach to a surface, head down.

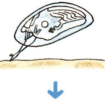
Settling. A larva finds a suitable spot, often near adults of its kind.

Gluing. The larva glues itself to the surface, head down.

Plating. It forms six lateral shell plates and a two-part lid.

Growing. As the barnacle matures, its shell grows larger.

How a barnacle's shell grows

When it first attaches to a surface, a second-stage barnacle larva forms six overlapping shell plates, arranged around it in a cone *(far right)*. A growing barnacle must occasionally shed the tough skeleton that encases the soft body within the hard shell. The shell, however, is never shed, and the shell plates grow continuously.

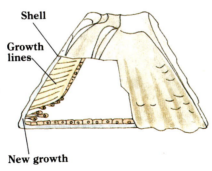

Shell

Growth lines

New growth

Adding shell. The growing barnacle normally adds layers of calcium carbonate—limestone—to the inside and the bottom and side edges of each shell plate.

Growing round. As its plates grow, the shell becomes higher and bigger around.

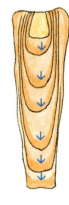

Growing up. A crowded barnacle extends the bottom of each plate, making a tall shell.

Why Do Lemmings Jump off Cliffs?

An old legend has it that colonies of lemmings leap into the far northern Atlantic Ocean in search of their ancestral home—the mythical sunken continent of Atlantis. The truth is less romantic—the lemmings are just migrating.

Lemmings reproduce rapidly and their populations can rise dramatically from year to year. Occasionally, when a population becomes too large, these 6-inch rodents leave their home territories on mass migrations in search of greener grass. If a stream bars their way, the whole crowd of lemmings just dives in and swims to the other side. But if they jump into a large river, a saltwater bay, or the ocean itself, the exhausted lemmings soon drown.

A lemming population migrates

Overpopulation. Although limited food usually controls their numbers, populations peak every three to six years. When they multiply to 25 times normal levels, lemmings strip the land bare, then seek new terrain.

Migration. The lemmings eventually settle new habitat, though many fall prey to hawks, foxes, and snowy owls along the way. Others perish when they reach impassable barriers such as rivers and sea cliffs.

Panic. A migration may begin as a trickle, but streams or cliffs can channel the lemmings into a torrent of panic-stricken rodents.

The north coast of Norway

Within the range of the Norway lemming (in purple on the map) lies the country's long coast, gouged by glaciers into a filigree of narrow bays called fjords. When thousands of the rodents migrate, some head for the coast. Like cowboys caught in a box canyon, huge migrations may find themselves atop a cliff with ocean on three sides. They pour over the cliff in their haste and panic—or perhaps because they see the water as just one more obstacle to be crossed. The geography of the Norwegian coast, then, may lie at the root of this strange behavior that has sometimes been mistaken for mass suicide.

Scandinavia

Lemming population peaks

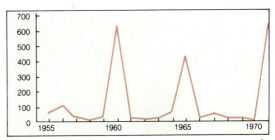

Alaska brown lemming populations, as measured by traps placed along their travel routes, peak about every five years, as shown above.

Why Are There Always Dead Trees in the Forest?

Like all living things, trees die of old age and disease. Other factors such as lightning, insects, and fire also take their toll. When a large tree crashes to the ground, it touches off a chain of events in the forest community.

The falling tree cuts a swath through the branches of its neighbors, allowing precious sunlight to stream to the forest floor. As the fallen trunk and branches rot, they add nutrients to the soil. Small trees use these nutrients and the sun's light for energy for a growth spurt, then shade their near neighbors and suck up water and minerals through their growing root systems. The slower-growing saplings eventually die for lack of light and water. When they in turn topple, they add fuel to the cycle of growth.

Competition in a forest

A fallen giant. A large tree has toppled, bringing sunlight to the forest floor.

A new stand. The slowest trees have died and healthy trees shade the forest floor.

Winners and losers. As crowding increases, slower-growing trees get less light, grow even slower, and die.

New giants. The patch now contains only a few mature, growing trees. As they die and fall in their turn, the cycle will be repeated.

A growth spurt. The sun's light and warmth spur tree seedlings to new growth.

Competition. As seedlings become saplings, they begin to compete for space and light.

What Is a 17-Year Cicada?

Cicadas are large, noisy relatives of leafhoppers. Often mistakenly called locusts, some cicadas spend 17 years as underground grubs or larvae. After that long childhood, thousands of these inch-long insects emerge all at once as adults to breed and lay eggs in tree branches. After the eggs hatch, the larvae drop to the ground and burrow into it, vanishing for this stage of life.

Most cicada species spend at least four years as underground larvae. But only three species of 13-year cicadas and three of 17-year cicadas, all found in the eastern United States, emerge simultaneously in May or June. The other species are dubbed dog-day cicadas because they appear during the hot, humid dog days of late summer; some of these adults emerge every year. Periodical cicadas may have evolved their synchronized breeding in response to predation.

Life cycle of 17-year cicadas

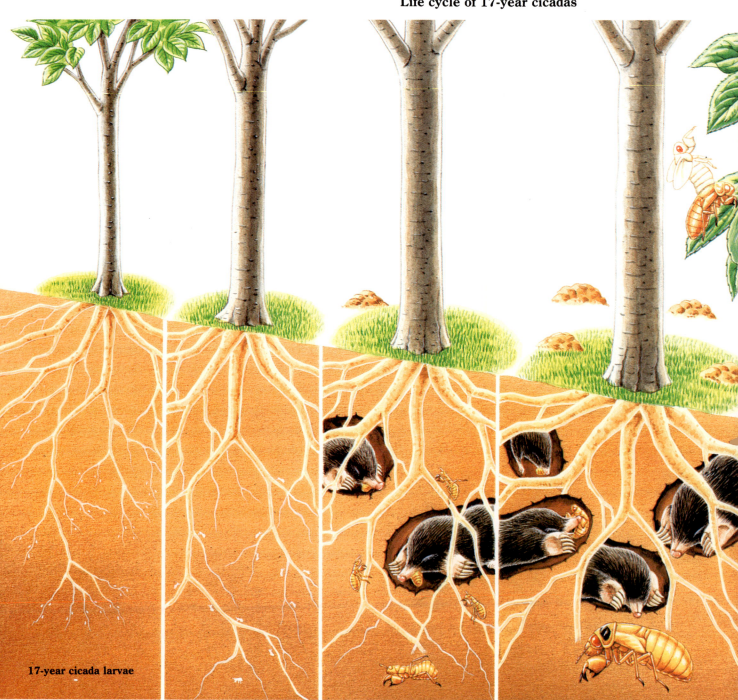

17-year cicada larvae

First year. First-stage cicada larvae the size of rice grains suck juices from tree roots.

Fifth year. Now in their second stage, the larvae have grown only a little larger.

Thirteenth year. Now the larvae are big enough to make a tasty morsel for a mole.

Sixteenth year. Mole populations may increase as they feed on fat larvae.

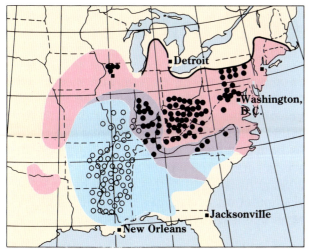

Distribution of cicada broods

The cicadas of one species that emerge from the ground in a given year are called a brood. Scientists have identified 13 broods of 17-year cicadas and five of 13-year cicadas, each brood emerging during a different year and living in a different geographic area.

This map charts two broods—the 13-year cicadas that came above ground in 1989 *(circles)* and the 17-year cicadas that emerged in 1991 *(black dots)*. While 17-year cicadas *(pink area)* generally occur in the North and the 13-year variety *(blue area)* in the South, their ranges do overlap *(purple area)*. Both kinds can inhabit the same woods but would emerge together only once every 221 years.

Distribution of
17-year cicadas

Distribution of
13-year cicadas

● 17-year cicadas
emerged in 1991

○ 13-year cicadas
emerged in 1989

Concurrent appearance of all three species

Seventeenth year—June. The larvae stage a mass exit from the soil and climb into the trees, where they metamorphose rapidly. Birds may feast on them but only for a few weeks.

Seventeenth year—July. After breeding and laying eggs in tiny slits they cut into twigs, the adults die. With their food source gone, the birds move on, leaving the tiny new larvae to drop to the ground, beginning the cycle once again.

Do Animals Control Their Numbers?

When animals reproduce, they try to maximize the number of healthy young in the next generation. In many species, this means producing as many young as possible, even if that leads to local overpopulation. But others take a different approach. Some insects have fewer offspring when conditions are crowded, thereby leveling out the year-to-year population changes in an area.

A female Japanese ladybird beetle, which feeds on thistles, will lay eggs only on unoccupied plants *(below, left)*, giving its eggs the greatest chance of maturing with plenty of food. But a female that doesn't find a suitable laying site reabsorbs its unlaid eggs, and the extra nutrition allows it to survive the winter and breed again the next year. As a result of this choosy behavior, a local population of Japanese ladybird beetles changes only slightly from year to year.

How Japanese ladybird beetles avoid population explosions

Searching. A ladybird flies to a thistle to lay its eggs.

Warned off. Seeing bare patches left by a feeding female, the ladybird flies away.

Giving up. Seeing another eaten leaf, it quits its search.

A female lays eggs on an untouched leaf.

Helping young survive

Every species has evolved special behavior to boost the chances that its young will survive to reproduce. Many insects lay eggs on what will be the offspring's food supply. Some leave odor marks to say, "I've already laid eggs here." A female of the same species wisely refuses to lay eggs near such a scent; if it did, its offspring would have stiff competition for food.

Seed beetles lay eggs on seeds, which the young bore into and eat. To ward off other seed beetles, a female marks its seed with scent.

Many wasps lay eggs in the egg case of other insects; the young eat the host. Some also scent-mark the victim to reduce competition.

The following spring

Wintering. The nutrients in the unlaid eggs will help it to survive the winter.

Having laid its eggs, it dies.

The female that hibernated lays eggs in the spring.

What Determines the Feeding Order Among Insects?

Sap oozing from the bark of a chestnut oak represents a nutritious feast to a wide variety of insects. Beetles, wasps, and butterflies all enjoy a sip. If there is not enough to go around, the insects will take turns, feeding in a definite order.

Among day-flying insects, strong scarab beetles command first place in line. Wasps follow, and butterflies and other beetles make do with leftovers. At night, a different group visits the tree, but again beetles lead the lineup. Rhinoceros and stag beetles fly in to feed, and male beetles often mate with females at the sap site.

A web of competition among insects feeding on sap

Sipping order. Many insect species vie for the sap. Certain species feed first; others follow.

Nymphalid

Rhinoceros beetle

Scarab

Wasp

Moth

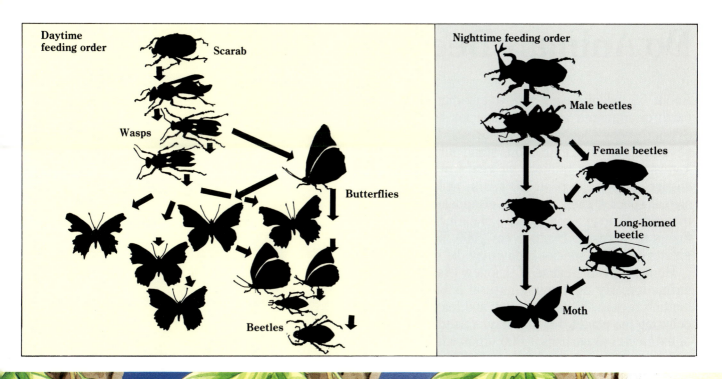

Daytime feeding order

Scarab

Wasps

Butterflies

Beetles

Nighttime feeding order

Male beetles

Female beetles

Long-horned beetle

Moth

Male stag beetle

Female rhinoceros beetle

Rhinoceros beetles

A male stag beetle waits for females.

● **Weapons of male beetles**

Among rhinoceros beetles, the males have enlarged mandibles, or hornlike appendages. The males use the mandibles as a weapon, a threat, or to attract and mate with the greatest number of females. As with deer, female beetles need no horns since they don't fight among themselves or with the males.

Do Animals Become Extinct Nowadays?

About 65 million years ago, every dinosaur on earth disappeared—with one species becoming extinct every thousand years or so. That and several other mass extinctions stand out in a geological history of millions of years of gradual change. But in the past 100 years, at least one animal species each year has disappeared from the earth forever. And the rate of extinction continues to increase. By the year 2000, some scientists say, 100 species a day may die out.

Humans are the cause of this new era of mass extinction. By destroying natural habitats, by hunting recklessly for food, fur, and feathers, by polluting the planet, humans have raised the rate of extinction to perhaps 1,000 times what it was before people appeared on earth. These pages show some victims and near-victims.

Passenger pigeon

There once were more North American passenger pigeons than any other species of bird. But hunters killed up to a million birds in a single season; the destruction of America's eastern forests further reduced their numbers. The last passenger pigeon, Martha, died in the Cincinnati Zoo in 1914.

Blue whale

Blue whales outweigh the largest dinosaurs and swim the deepest oceans. But with new technology to locate and harpoon the whales, people have brought the great blues close to extinction. International protection laws may help blue whales stage a comeback.

Endangered species

When a population becomes so small that extinction becomes a possibility, that species is put on the endangered list. Populations shrink for many reasons, and some species have never achieved large populations. Even moderate hunting or disease can damage a species. Other species need a particular habitat or a large range. If that habitat shrinks, the species, too, may dwindle.

Mountain gorilla. This largest of all primates has always been scarce. Now, poaching and habitat loss have slashed its numbers.

Giant panda. Because of hunting and capture for zoos, the wild giant panda now lives only in the bamboo forests of central China.

Cheetah. The world's fastest land animal is easily tamed, but wild populations plummet when people move near their terrain.

What Happens When a Fish Dies?

After a fish dies, its body slowly sinks toward the ocean floor. If it is not devoured by sharks on the way down, the dead animal will touch bottom and fall prey to a host of hungry scavengers—animals that feed on carrion or refuse.

First to feast on the carcass are larger, more mobile species such as scorpionfish, crabs, and lobsters. Next on the scene are smaller, slower creatures such as hermit crabs, starfish, snails, and shrimp, which consume most of the flesh.

Any remains are attacked by bacteria. These microscopic organisms break down the flesh into simple organic and inorganic materials. Within one or two weeks, the dead fish has been reduced to a skeleton—and that, too, may eventually decompose beyond recognition.

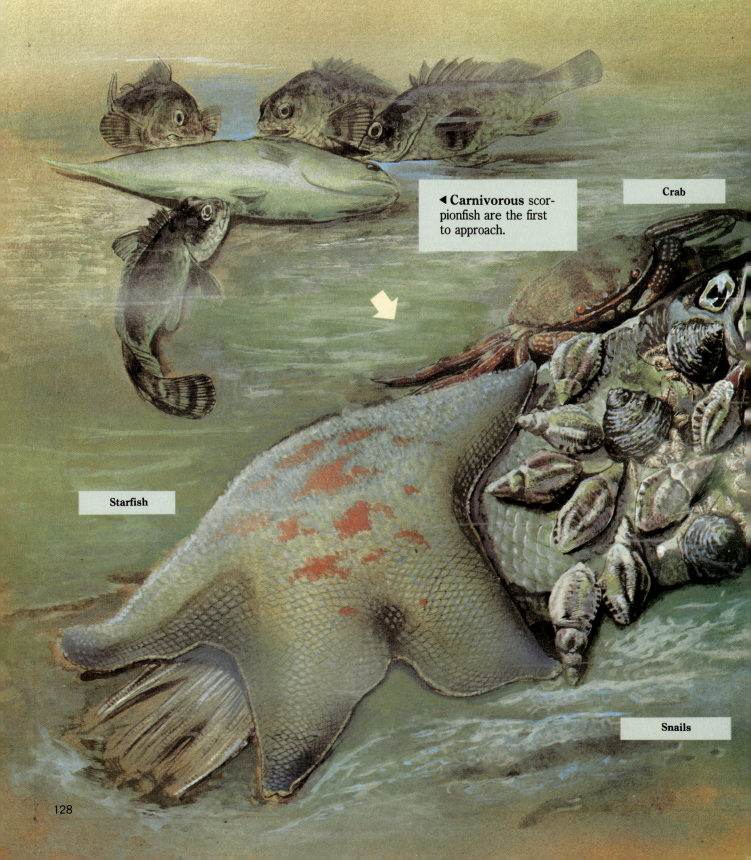

◄ **Carnivorous** scorpionfish are the first to approach.

Crab

Starfish

Snails

128

One species saved

Like many large birds, the graceful whooping crane was once a favorite target of hunters. Besides hunting, the conversion of whooping crane habitat to farmland also took its toll; by 1941, only a single flock of 15 birds remained. But vigorous protection and the help of scientists—using sandhill cranes as nannies—have increased the flock to a little more than 100, a fraction of the whooping crane population before hunting began.

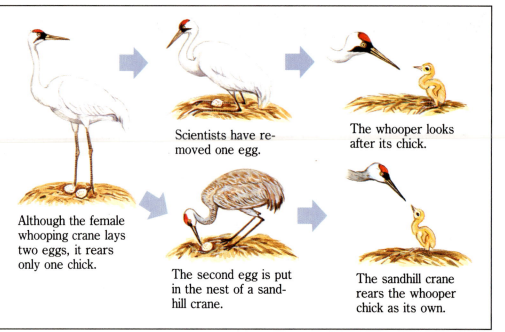

Although the female whooping crane lays two eggs, it rears only one chick.

Scientists have removed one egg.

The whooper looks after its chick.

The second egg is put in the nest of a sandhill crane.

The sandhill crane rears the whooper chick as its own.

Land iguana

Land iguana populations have suffered the twin attacks of hunting and habitat destruction. Both humans and their cats kill iguanas for food, and on some islands, wild goats have destroyed the vegetation, leaving no hiding places. Once plentiful on the Galápagos Islands, these 2-foot-long lizards have been reduced to small, isolated populations.

African elephant

The animal with the world's largest ears and longest front teeth, the African elephant, has been hunted to near extinction in some parts of Africa. Those front teeth, or tusks—which can reach 11 feet 6 inches—provide precious ivory. Though the elephant is protected in game reserves and national parks, poachers still take a toll each year.

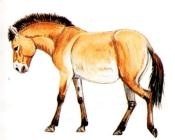

Wild horse. Hunting and breeding with domestic horses have nearly wiped out the Przewalski's horse of southwestern China.

Japanese crested ibis. This pinkish white bird has become nearly extinct because of pollution and the loss of its habitat.

Javan rhinoceros. While all rhinoceros species are endangered, the Javan rhinoceros is closest to extinction.

Galápagos tortoise. Like the land iguana, the Galápagos tortoise has suffered from being hunted by humans and domestic animals.

What Animals Disperse Pollen?

Many plants could not reproduce without help from animals that carry pollen from one flower to another, enabling the plants to set seeds. Most pollen-dispersing animals fly or glide from one plant to another in their search for food, carrying pollen only by accident. The largest group of pollinators is the insects, including bees, butterflies, and beetles. Many fruits and flowers rely on bees for pollination.

Hummingbirds are important pollinators of day-blooming flowers in North, Central, and South America, while fruit bats serve the night-blooming flowers throughout the tropics. In Australia, a bird called the white-cheeked honey eater fills a role like the hummingbird's, collecting nectar and carrying pollen. Another Australian, a marsupial called the pygmy glider *(bottom left)*, soars from tree to tree, feeding on nectar and pollen at night—thereby pollinating flowers—as well as gobbling insects.

Creatures that disperse pollen

Fruit bat

Malachite sunbird

White-cheeked honey eater

White-eye

Hermit hummingbird

Pygmy glider

Flowers and pollinators

Many animal-pollinated plants, such as *Centropogon*, a member of the Lobelia family *(right)*, have evolved tubular flowers with nectar deep within and pollen-laden stamens near the entry. A bird reaching for the nectar brushes the stamens, picking up pollen. The pollinators have evolved beaks that are suited to narrow blooms.

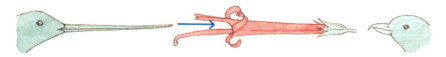

The beak of the Magnificent hummingbird reaches deep into *Centropogon*'s bloom.

The nectar of *Centropogon* lies inside the base of the elongated flower.

The slaty flower-piercer punctures the flower's base with its short, sharp beak.

Swallowtail butterfly

Bumblebee

Honeybee

Dronefly

Scarab

115

Why Do Some Plants Eat Insects?

Despite their reputation as voracious predators, insectivorous plants capture insects mainly to obtain nitrogen, not to satisfy a taste for blood. Plants need nitrogen and phosphorus in order to synthesize protein, but insectivorous plants live in areas such as bogs that are poor in these elements. These plants lure insects and small animals by their attractive appearance or smell.

By quick snaps and sticky traps they catch their prey, then digest it with acids and enzymes they secrete from glands. Thus reduced to a nutrient-rich liquid, the meal is absorbed into the plant by special cells. Insectivorous plants can survive without animal proteins in their diet, but they are healthier when they catch an occasional fly or crustacean.

The sundew plant lures insects with its sticky red hairs.

The modified leaf encloses the trapped insect.

Plant traps for insects

Aquatic bladderwort

Venus flytrap

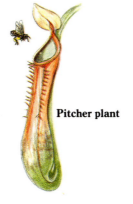

Pitcher plant

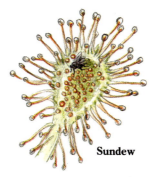

Sundew

Suction. When sprung, a one-way valve sucks in prey swimming close by.

Jaws. An insect touching the outer hairs triggers the trap.

Drowning pool. Hairs trap prey in a vat of water or digestive juices.

Stickiness. Sticky droplets on the tentacle tips work like flypaper.

Evolving traps for food

All plants need nitrogen and phosphorus; most get them from the soil through their roots. Insectivorous plants have adapted to poor soils and found another source for these elements. Over time, their root systems have shrunk, and their leaves have evolved elaborate ways to capture nutrients from animals. Like other plants, insectivorous plants make starch by photosynthesis.

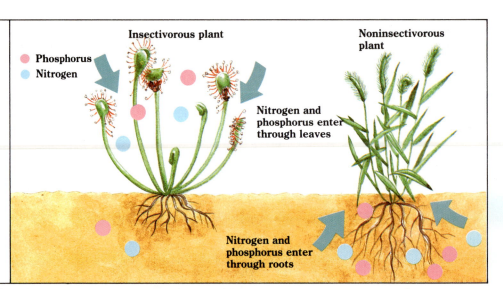

Insectivorous plant

Noninsectivorous plant

● Phosphorus
● Nitrogen

Nitrogen and phosphorus enter through leaves

Nitrogen and phosphorus enter through roots

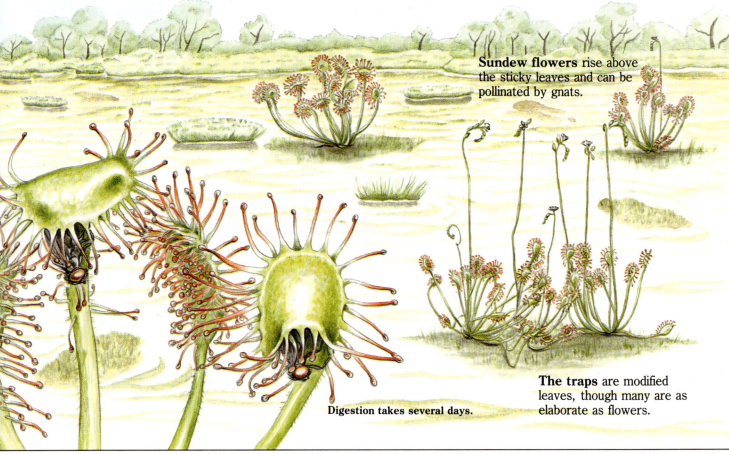

Sundew flowers rise above the sticky leaves and can be pollinated by gnats.

Digestion takes several days.

The traps are modified leaves, though many are as elaborate as flowers.

Digestion and absorption

The catch. Attracted by nectarlike droplets on the sundew's hair tips or by the redness of its tentacles, a bee gets stuck to the plant's tentacles.

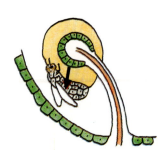

Enclosure. Having caught the prey, the leaf folds over on it. In about 10 minutes, the insect is surrounded by many sticky hairs.

Digestion. As the leaf bends, the insect touches acid-secreting glands on the leaf surface. Acid enters the body and begins to digest it.

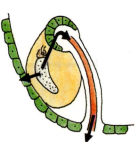

Absorption. The hairs and the leaf surface form a temporary pouch in which the prey is digested and its nutrients absorbed into the plant.

What Defenses Do Plants Have?

Since plants cannot escape from creatures that feed on them, many have evolved passive ways to discourage herbivorous, or plant-eating, creatures. The two main defenses plants use are armor and poison.

Anyone who has been pricked by a cactus or cut by razor grass has felt the armor. Spines, barbs, or stinging cells make such plants unpleasant to eat or touch. One variety of cactus *(opposite)* grows in a special shape to defend itself.

Poisonous plants produce chemicals with effects ranging from discomfort to death. Oils in poison oak, poison ivy, and stinging nettles cause an itchy rash. The leaves of tomato and potato plants are toxic to some insects and grazers. Still other plants produce such poisons as strychnine and cyanide; while animals may nibble on these plants once, they quickly learn to avoid them.

Deerproof. Japanese azaleas *(red, below)* produce toxic chemicals that let them survive even among hungry deer. Browsing deer avoid the azalea bushes, even after cropping all surrounding leaves and bushes.

Competitive edge. Even when browsers like these deer are numerous, poisonous plants flourish. After their tastier neighbors have been eaten, the toxic plants receive more of the light and nutrients that they need for growth.

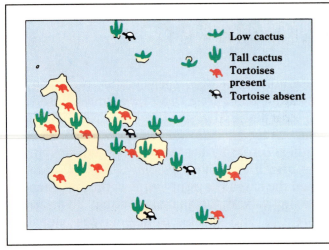

	Low cactus
	Tall cactus
	Tortoises present
	Tortoise absent

Specialized Galápagos cacti

In the Galápagos Islands, the prickly pear cactus has evolved into two distinct shapes, apparently in response to the feeding habits of the Galápagos tortoise. A low, creeping cactus with soft spines can be found on islands where there are no Galápagos tortoises. A treelike variety with a tall, thick trunk and hard, tough spines grows on the islands where the tortoise lives or has lived.

The cactus on the islands with tortoises seems to have reacted to the animal by evolving stronger defenses against it. Over time, the taller cactus plants survived better than the lower, softer plants, so that eventually all of the cactus plants on those islands were the taller and tougher kind.

A safe height. Cacti on islands inhabited by tortoises grow a thick stalk more than 6 feet tall, putting the tender leaves well out of the reach of the creatures.

Safety on the ground. On islands that never had tortoises, the cacti grow in ground-hugging clumps. The spines on the pads are apparently sufficient defense to allow the cactus to survive.

119

How Is a New Habitat Colonized?

In 1883, a volcanic eruption on the island of Krakatoa, Indonesia, destroyed two-thirds of the island and covered the rest with ash and pumice. The blast eradicated all life on the island. Nine months after the explosion, biologists found only a few kinds of grasses and a single spider. Since then, biologists have studied Krakatoa as a natural experiment in colonization. First, blue-green algae covered the shorelines. Then came tropical plants, such as coconut palms, which have seeds that can float in seawater for weeks without damage. Ferns and grasses, which have spores and seeds light enough to be carried by the gentlest air currents, colonized the island next. Soil began to form from decaying plant matter, providing a habitat for more plants and for animals. Insects and birds flew to the island or were carried by the wind. Other animals—including two species of reptiles—probably floated over on logs or rafts. The arrival of pollinators *(pages 114-115)* hastened the spread of plant life, in return providing a richer habitat for animals. Within 20 years, more than 260 species of insects and animals made their home on the island. Today the island has recovered with more than 1,200 species of animals established in a dense rain forest. But colonization is not complete—biologists continue to find new arrivals.

After the eruption, airborne algae spread along the shore to form a base for the germination of seeds. Flying insects, birds, and bats brought seeds along with them. Other seeds and small creatures, such as spiders *(top)*, were carried by the wind. Coconuts and other seawater-resistant seeds drifted in on ocean currents and took root. The jumping spider *(above, far left)* and blind snake *(above, left)* arrived on driftwood.

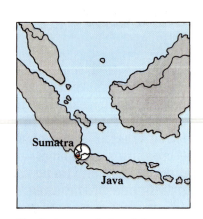

Krakatoa, a chain of three volcanoes that until 1883 formed a 17-square-mile island, is located between the Indonesian islands of Java and Sumatra. When the volcanoes erupted, the explosion blew two-thirds of the island away, leaving it barren. Today a new, smaller cone is rising up through the water from what was once the island's center.

Krakatoa today. The lush growth of a rain forest covers the island once again. As trees grew tall, many of the early pioneers, such as grasses, disappeared and were replaced by other species that had adapted to the dense and sunless rain forest. Monitor lizards *(above, left)* and pythons swam ashore, while geckos *(above, right)* and other lizards probably floated in on logs or clinging to debris.

How Do Animals Disperse Seeds?

Whenever an animal walks through a field or forest, flies from tree to tree, or eats a piece of fruit, it may be unknowingly helping to disperse plant seeds. Seeds can hitch rides on the outsides or insides of animals. Some seeds hook like Velcro onto an animal's fur *(below)*, and others use a gum that sticks to fur, feathers, and feet. A few seeds, like the trample-burrs of Madagascar, can do real damage when they attach themselves with their large, sharp thorns to an animal's feet.

Many plants get their seeds inside an animal by offering an attractive fruit in exchange for seed transport. When an animal eats the fruit, the tough-coated seeds pass unharmed through the animal's digestive tract. Later on, they will emerge in the creature's droppings—with any luck, in an appropriate habitat. Being carried on or inside animals is one of the most common ways in which many plant species disperse their seeds and colonize new areas.

Hooking a ride. The tickseed has hooks and hairs that fasten to animal fur.

Velcro. The cocklebur has barbed hooks that attach it securely to fur or clothing.

Glue. A gluey substance sticks the *Siegesbeckia pubescens* seeds to passersby.

Spines. Sharp spines help burr marigold seeds, known as beggar-ticks, hitch rides on animals.

Air mail. Birds eat many fruits and eliminate the seeds in their droppings, perhaps miles away.

Carrier. Squirrels disperse acorns and other nutlike seeds.

Ground transport. Mice and rats also collect seeds.

Planter. The Japanese jay buries seeds.

123

4

The Food Chain

Every living thing needs food in order to exist. Most plants are able to make their own. Through a process known as photosynthesis, they use the energy of the sun to convert water, carbon dioxide, and minerals into life-sustaining organic matter. As food producers, plants occupy the bottom level of the food chain—an arrangement of organisms in an ecological community in which each member feeds on the one below it.

Animals, which eat plants or other animals to obtain nourishment, are referred to as consumers and form the higher links of the chain. At the

top of the structure, which is often diagramed as a pyramid, is the ultimate consumer: a creature with no natural predators, such as the eagle, whose food chain is depicted above.

Since most animals eat a variety of foods, a food chain often overlaps and interweaves with several others. Indeed, the food chain is part of the overall cycle of growth and decay that characterizes life on earth. When a creature at the top of a food chain dies, bacteria eventually break it down into the simple organic and inorganic compounds that are used by plants to manufac-

ture food. Similarly, when plants die, they too are reduced by fungi and bacteria to organic and inorganic matter that will in turn fuel a new generation of plants.

Insects, toads, and snakes occupy the lower links of a food chain topped by an eagle, shown feeding its eaglets a morsel of snake flesh at top left. The illustration below portrays the eternal cycle of growth and decay; here, the remains of a fallen tree nourish a crop of saplings.

How Do Eagles Grow?

Golden eagle

Copper pheasant

A majestic bird of prey with a wingspan of up to 6½ feet, the golden eagle feeds mostly on rabbits, snakes, and large game birds, such as pheasants. In February or March, the female normally lays two eggs. After 43 to 45 days of incubation, the first chick breaks free of its shell; the second chick follows in three or four days. Soon after the second chick hatches, its older—and larger—sibling may kill it by keeping it from feeding, pushing it out of the nest, or even pecking it to death. This deadly aggressive behavior, which occurs most often when food is scarce, seems to have evolved as a method of ensuring that at least one chick in each brood will survive. When prey is plentiful, the aggression subsides and both chicks are raised to maturity.

Jack rabbit

Snake

Fly

Frog

Grasshopper

The route to becoming a raptor

A golden eagle chick sits by its unhatched sibling. The mother lays its first egg three or four days before the second; the eggs hatch in that order.

An older chick pecks at its smaller and weaker nestmate. Even when death is imminent, the parents make no attempt to separate their dueling offspring.

Two young golden eagles share a nest. When food is plentiful, the older chick usually allows the younger one to live.

The golden eagle food chain

Golden eagle

Snakes

Frogs, toads, lizards

Flies, locusts, other insects

The golden eagle occupies the top of its food chain, shown above as a pyramid graph. Among the bird's prey are snakes, which feed on frogs, toads, and lizards; these in turn eat flies and other insects. As the graph indicates, the prey far outnumber the predators.

Scavengers of the sea

Different scavengers inhabit various depths of the ocean. A dead fish that falls to the bottom of a sandy coastal area, for example, will be consumed by thick-bearded shrimp, lugworms, clamworms, ribbon worms, hermit crabs, and snails. In deeper waters, the carcass will be devoured by fish and a larger species of shrimp. Depending on the ocean currents, scavengers are often able to scent carrion from a considerable distance.

▲ Lured by the smell, hermit crabs and snails feed on the flesh of a fish that has been dead for two days.

▲ Because they attract scavengers, dead fish are often used as bait in crab pots *(above)* and other traps and nets.

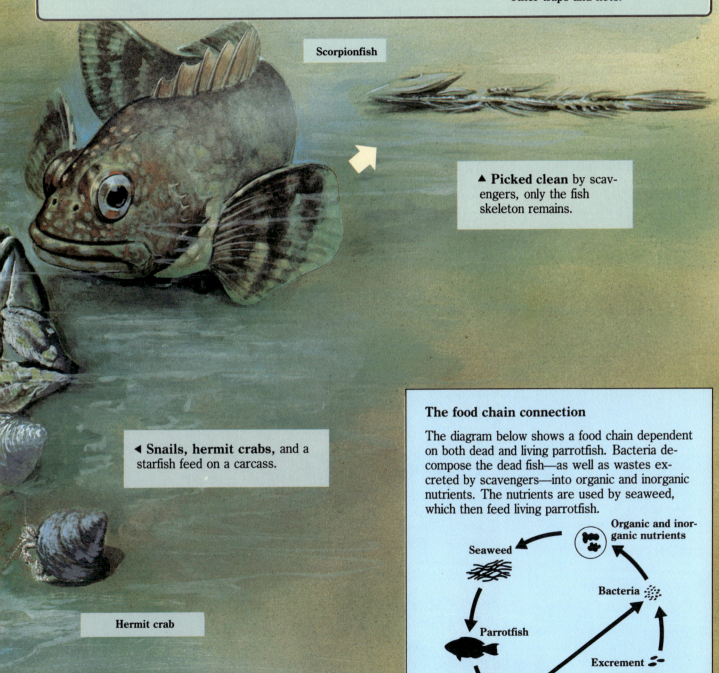

Scorpionfish

▲ **Picked clean** by scavengers, only the fish skeleton remains.

◄ **Snails, hermit crabs,** and a starfish feed on a carcass.

Hermit crab

The food chain connection

The diagram below shows a food chain dependent on both dead and living parrotfish. Bacteria decompose the dead fish—as well as wastes excreted by scavengers—into organic and inorganic nutrients. The nutrients are used by seaweed, which then feed living parrotfish.

Organic and inorganic nutrients

Seaweed

Bacteria

Parrotfish

Excrement

Dead fish

Scavengers

How Do Sea Otters Live?

The sea otter, a large member of the weasel family, is known as a keystone species—an animal that has a marked impact on its habitat. The extent of the otter's influence can be demonstrated by comparing the coastal regions of two groups of islands in the Aleutian chain. In the Near Islands, where only a handful of sea otters live, the ocean floor has sparse patches of seaweed but dense clusters of sea urchins. Relatively few fish are to be found here.

The Rat Islands, by contrast, have a sizable sea otter population, and the large brown seaweed called kelp flourishes in great groves. These floating forests support a wide variety of marine life, including multitudes of fish.

The key difference between the two ecosystems is the absence or presence of the sea otter: It feeds on sea urchins, which eat kelp. By removing the sea urchins, the otter enables the kelp—and the fish with it—to grow and thrive.

The impact of the sea otter

The diagram below shows the sea otter's influence on the coastal food chain. When otters are present, kelp thrives, as do the shrimp and crab larvae that feed on it or hide among it. The fish that eat these tiny crustaceans also abound, luring seals to the region. In seas without otters, the proliferation of the kelp-eating sea urchins alters the food chain. Bottom dwellers such as mussels and starfish flourish, while the fish population declines. With fewer predators, more shrimp and crabs mature, attracting octopuses to the area.

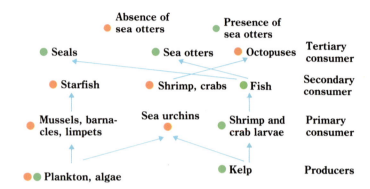

● Absence of sea otters		● Presence of sea otters	
● Seals	● Sea otters	● Octopuses	Tertiary consumer
● Starfish	● Shrimp, crabs	● Fish	Secondary consumer
● Mussels, barnacles, limpets	Sea urchins ●	● Shrimp and crab larvae	Primary consumer
●● Plankton, algae		● Kelp	Producers

■ An ocean with no otters

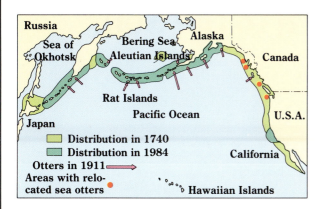

Russia
Sea of Okhotsk
Bering Sea
Aleutian Islands
Alaska
Canada
Rat Islands
Pacific Ocean
U.S.A.
Japan
☐ Distribution in 1740
☐ Distribution in 1984
Otters in 1911 ⟶
Areas with relocated sea otters ●
California
Hawaiian Islands

An otterly successful relocation

Explorers in the early 1700s found the coastal regions of the northern Pacific Ocean teeming with sea otters, as indicated by the light- and dark-green shading on the map at left. A frenzy of hunting and fur trading ensued, and by 1911, the sea otter was on the verge of extinction; it survived only in those locations marked by red arrows. In that year, a group of nations agreed to protect the creature from further hunting. The distribution of the species was also expanded by the successful relocation of otters to new colonies *(red dots)*. By 1984, thanks to the efforts of conservationists, sea otters once again thrived in much of the northern Pacific *(dark-green shading)*.

■ **An ocean filled with otters**

How Does a Tree Return to the Soil?

Death and decay of a hemlock

Tree seedlings

Fallen hemlock tree

1 Bacteria and fungi that enter a tree through its broken branches or damaged roots eventually cause the tree to weaken and collapse, often with its bark and branches intact. Although the hemlock tree above appeared healthy when cut, its core had begun to decay from a fungal disease known as brown rot.

▲ Invaded by fungi, the core of the tree begins to turn black.

2 After 5 to 10 years, the sapwood, or outer section, starts to turn white from decay, while the heartwood continues to deteriorate. The branches have fallen off and the bark begins to decompose. The decay is most advanced where the tree is moistened by the soil.

▲ Fungi secrete enzymes that gradually break down the cell walls of the wood.

Not long after it hits the ground, a fallen tree is attacked by hordes of creatures from the soil. The first onslaught generally comes from grubs, or beetle larvae, which begin to bore and eat into the dead wood. Termites, beetles, ants, slugs, and snails form the next wave of assault, tearing the timber into tiny fragments as they feed.

Soon, all sorts of fungi—the chief agents of forest decomposition—invade the tree or attach themselves to it. Fungi are scavenger plants that grow on dead plant matter; they cause wood and bark to soften and rot. Not only does the decay make the tree more vulnerable to attack by insects and other animals, but it also creates a welcoming environment for the bacteria that hasten the process of decomposition. Certain fungi are even able to break down lignin, the bacteria-resistant substance that makes up a tree's central section, or heartwood. Over a period of between 20 and 100 years, the forces of decomposition reduce the fallen tree to a spongy and extremely nutritive substance called humus, which is eventually worked into the soil by earthworms and other creatures.

3 Ten to 20 years later, almost all the heartwood has rotted away; only the outer shell and a little bark remain. The timber has become waterlogged and highly acidic. Moss and ferns, which thrive in such conditions, begin growing over the top. Tree seedlings—deposited by birds or the wind—start to climb skyward.

4 In 20 to 40 years, the trunk has collapsed upon its core. Next it will begin to crumble into chunks, which in turn will decompose into humus—the dark organic material that makes soils so fertile. The seedlings, thinned out by competition for sunlight and fed by nutrients from the fallen tree, have grown into saplings.

5 Fifty to 100 years after falling, the hemlock has decomposed entirely. The organic material that the tree once contained has returned to the soil in the form of humus. The humus nourishes a new generation of trees, which neatly demarcate the remains of the dead hemlock.

How Do Mushrooms Grow?

■ **A forest full of fungi**

Oak tree

Bearded hedgehog

Brickcap mushroom

Caesar's amanita

Turkeytail

Coincap
(Fairy helmet)

Webcap

Cordyceps

▼ In a mycorrhiza, the mushroom's hyphae are interwoven with tree roots.

There is more to the mushroom than meets the eye. The part that grows from the ground or on trees is only the short-lived fruiting body of the plant. The rest of the mushroom, known as the mycelium, lies hidden beneath the surface. Composed of a tangle of rootlike structures called hyphae, the mycelium takes in nourishment for the mushroom plant.

Most mushrooms subsist on dead plant matter—rotting wood and decaying leaves, for example—which they further break down, or decompose, into simpler forms. A few varieties, however, feed on living plants. Some of these parasitic species have mycelia that kill the host plant by causing it to rot. Others are less preda-

tory; they exchange nutrients with their host through a mycorrhiza, a symbiotic union of the mycelium and the roots of the plant. Parasitic cordyceps mushrooms *(far left)* feed not on plants but on animals: Their mycelia grow on the bodies of insects in the larval or pupal stage.

All mushrooms begin as microscopic single cells called spores. In gilled mushrooms, millions of spores develop on the gills in club-shaped bodies known as basidia. Ripe spores are shot out of the basidia and dispersed by the wind and by animals. If a spore lands in a place with suitable nutrients, it begins to grow. And if the spore's mycelium joins with that of another spore, the mushroom will reproduce, as diagramed below.

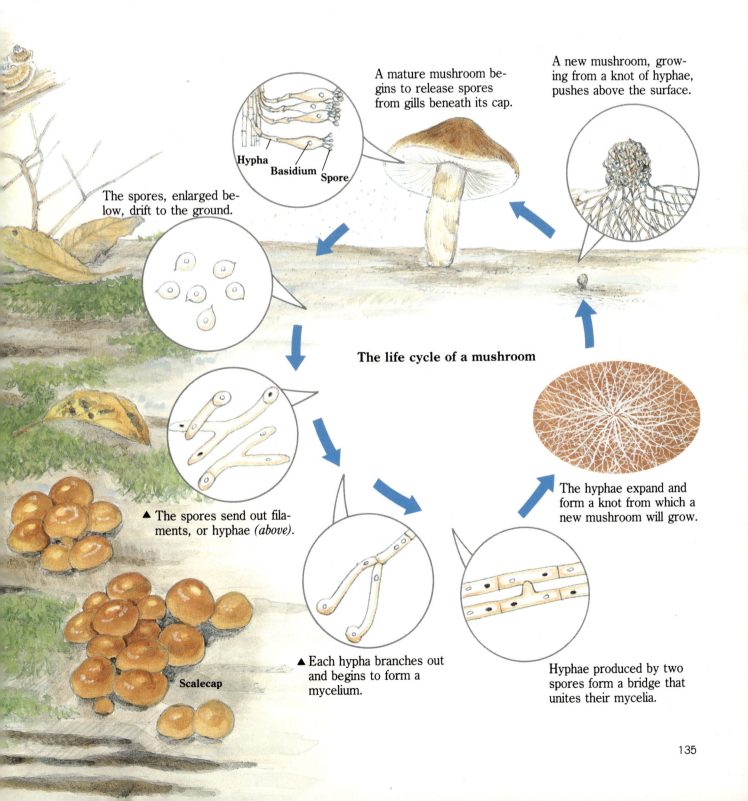

A mature mushroom begins to release spores from gills beneath its cap.

A new mushroom, growing from a knot of hyphae, pushes above the surface.

Hypha   **Basidium**   **Spore**

The spores, enlarged below, drift to the ground.

The life cycle of a mushroom

▲ The spores send out filaments, or hyphae *(above)*.

The hyphae expand and form a knot from which a new mushroom will grow.

Scalecap

▲ Each hypha branches out and begins to form a mycelium.

Hyphae produced by two spores form a bridge that unites their mycelia.

Can a Forest Recover from a Fire?

At first glance, a forest ravaged by fire might seem absolutely dead. But a closer look will reveal signs of new life. Ecologists have found that fires pave the way for a rebirth, rather than marking the end of a forest. As fire destroys the dense, sun-blocking forest canopy, new trees and plants can take root on the sun-warmed forest floor. With the first seed, the cycle of regrowth and species interdependence begins again. Insects invade the dead trees, followed by birds and small animals that prey on them. New plants attract the plant-eating species. The nutrient-rich ash from the fire falls into rivers, nourishing microbes and algae, which provide food for fish, which in turn are eaten by bears and other mammals.

In 1988, for example, a series of fires roared through nearly one million acres of pristine forest in Yellowstone National Park, leaving much of the park a bleak, ashen landscape. By the following spring, so many new plants had sprouted and so many animals had returned that it was hard to believe that the land had been completely barren a few months earlier.

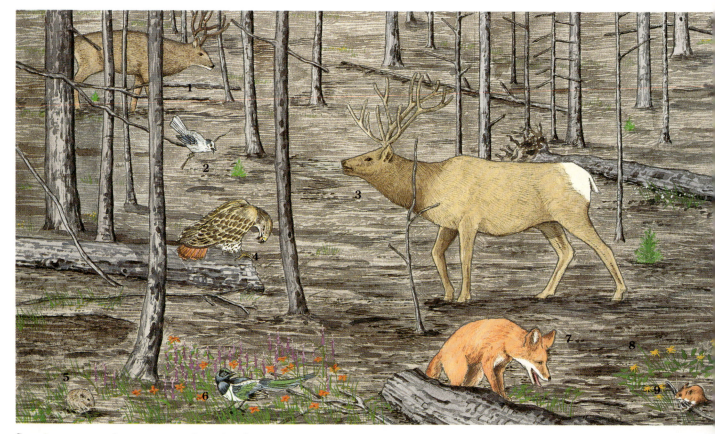

Species return three years after the fire

1 Mule deer	3 Elk	5 Vole	7 Red fox	9 Deer mouse
2 Clark's nutcracker	4 Red-tailed hawk	6 Black-billed magpie	8 Arnica	10 Black bear

The forest's cycle

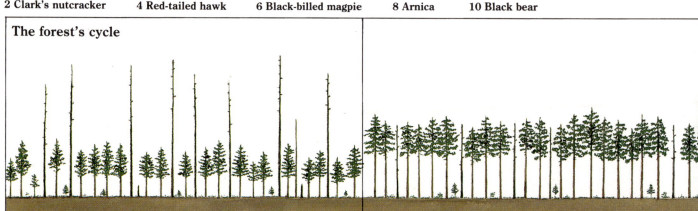

After a fire, many new trees take root. Lodgepole pine cones, for example, release seeds in the heat of fire.

In 50 to 150 years, lodgepole pines reach maturity, forming a canopy 30 to 50 feet above ground.

The debate over fighting fires

Some ecologists want to preserve the forests as they are and advocate putting out all fires immediately. But as the role of forest fires is becoming clearer, that attitude is changing. Today many ecologists prefer putting out only those fires that were caused by human carelessness. Fires started by lightning are allowed to burn themselves out if they pose no threat to human lives or developed property. By letting some fires run their course, ecologists hope to maintain a forest's mosaic—its patchwork pattern of old and new areas. Such a mosaic allows for the regeneration of some forest sections. Young, green trees in the newer sections help prevent the destruction of the entire forest when fires strike.

11 Mountain bluebird   13 Chipmunk   15 Blue grouse   17 Yellow-bellied marmot
12 Lodgepole pine cone   14 Asters   16 Purple fireweed   18 Coyote

After about 150 years, dying lodgepole pines create gaps in the canopy. Spruce and fir begin to flourish.

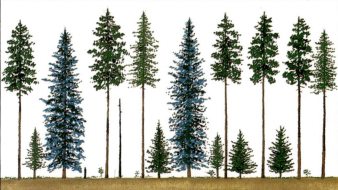

After 300 years, the number of old, dry trees and fallen branches make the forest once again vulnerable to fire.

How Do Alligators Help Other Wetlands Creatures in the Dry Season?

Wetlands are delicate buffer zones where land and water meet; some are swamps, or forested wetlands; others are fresh- and saltwater marshes, or grassy wetlands. With their soggy soil, wetlands were once thought of as mosquito-infested wastelands and were often dredged and filled to make room for new housing developments. Today wetlands are recognized as some of the most vibrant habitats in the world.

Among wetlands, the alligator holes of the Florida Everglades fill a special ecological niche.

During the dry season—in winter—these fresh-water ponds are often the only watering holes for Everglades creatures.

Alligators wallowing in shallow ponds clear the bottoms of plants and muck and, with blows of their tails, keep the ponds from silting over. Once an alligator has excavated a hole to its liking, it will use the hole year after year. Fresh-water shrimp, gar, killifish, snails, and snapping and soft-shelled turtles live alongside the alligators year-round. As the water level drops in the

At dawn and dusk many animals come to drink and feed at an alligator hole, shown symbolically here. The watering hole links these creatures in a complex food web. The simplest life forms, the algae and single-celled organisms, provide food for mosquito and frog larvae. In turn the larvae are eaten by small fish, such as gambusia and killifish. The small fish are prey for larger fish and also for wading birds, mammals, and reptiles.

surrounding marsh, the holes become crowded with larger marsh-dwelling fish and attract otters, raccoons, and wading birds. The soft mud banks support lush vegetation that provides for herbivores, or plant-eating animals, which then become prey for carnivores, or flesh-eating animals.

The crowd at the alligator hole

1 Black-crowned night heron	14 Pig frog
2 Florida panther	15 American egret
3 Roseate spoonbill	16 White-tailed deer
4 Green tree frog	17 Snowy egret
5 Alligator	18 Snail kite
6 Florida spotted gar	19 Apple snail
7 Mosquito	20 Tree snail
8 Gambusia	21 Mockingbird
9 Largemouth bass	22 Little blue heron
10 Soft-shelled turtle	23 Marsh rabbit
11 Bluegill	24 Box turtle
12 Killifish	25 Great blue heron
13 Purple gallinule	26 River otter
	27 Raccoon

What Is an Estuary?

An estuary is a semienclosed, coastal body of water, where fresh water from rivers mixes with salt water from the ocean. A typical estuary and the largest in the United States is the Chesapeake Bay on the East Coast. The bay and its tributaries are surrounded by extensive marshland, which provides a habitat for a multitude of birds, reptiles, and mammals. The wetlands grasses enhance the water quality by trapping sediments and absorbing nutrients. When the plants die back in the winter, their leaves and stems break down into small particles, called detritus, which are flushed into the water and become part of the food chain. Because the bay is only 21 feet deep on average, sunlight promotes the growth of vast beds of rooted aquatic plants. When these sea grasses decompose, they also form detritus. The detritus, together with microscopic floating plants, called phytoplankton, makes up the base of the bay's food chain.

Although food is plentiful, fluctuating levels of salinity and temperature make the bay a hard place to live. Only a few species, including the oyster and the blue crab, are able to use the bay as their year-round home. This limited diversity has resulted in large populations of the species that do live there. The various bay inhabitants can be found in distinct communities, such as sea-grass beds or oyster bars, depending on water depth and salinity zone.

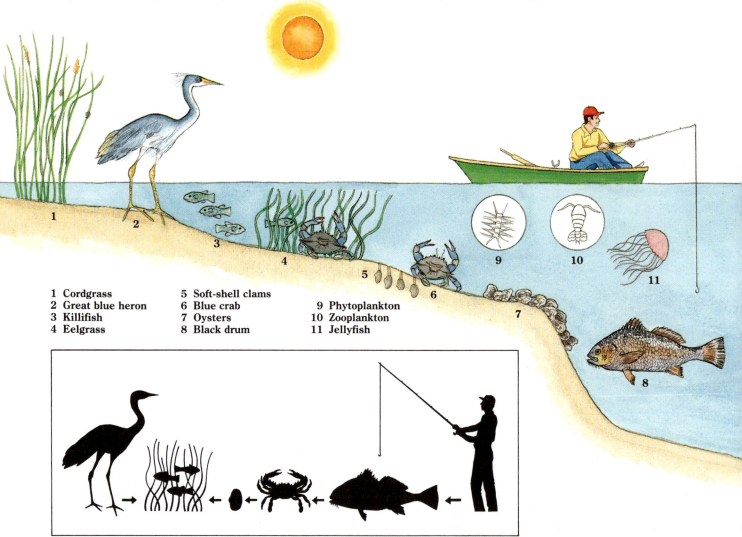

1 Cordgrass
2 Great blue heron
3 Killifish
4 Eelgrass
5 Soft-shell clams
6 Blue crab
7 Oysters
8 Black drum
9 Phytoplankton
10 Zooplankton
11 Jellyfish

Life in the bay

The diagram above depicts two of the many strands that make up the estuary's food web. But the interaction is more complex. Cordgrass lines the marshes, where the great blue heron stalks killifish, which seek shelter and food in the eelgrass. Decomposing grasses from the marshes release nutrients that feed phytoplankton, which in turn are eaten by zooplankton. Plankton is also a food source for jellyfish, oysters, and small crustaceans. Crustaceans, such as amphipods, are consumed by killifish and other small fish. The marsh detritus also nourishes soft-shell clams, which are preyed upon by blue crabs. The blue crabs are eaten by the black drum that feed on the bottom. And at the top of the food chain are humans, who make meals of oysters, clams, crabs, and fish.

Estuarine species

Stretching between Maryland and Virginia, the Chesapeake Bay (rectangle at left and below) and its tributaries are surrounded by nearly 500,000 acres of marshland. Of the 295 species of fish that live in the bay during the summer, only 29 species are able to tolerate the Chesapeake's fluctuating salinity and stay year-round.

Largemouth bass *(above)* live in fresh water but also range into the slightly brackish water in the upper parts of the Chesapeake Bay and its tributaries. Some 46 kinds of freshwater fish inhabit this area, with an additional 32 species straying to these feeding grounds on occasion.

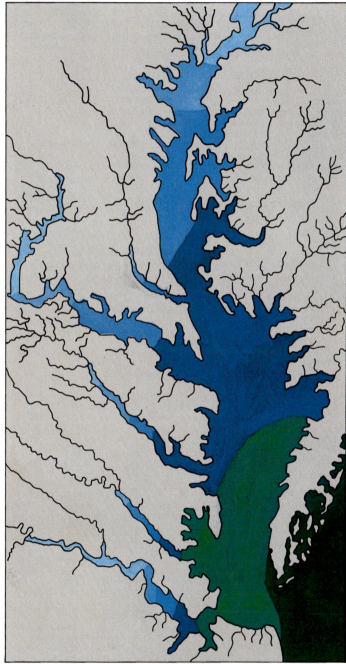

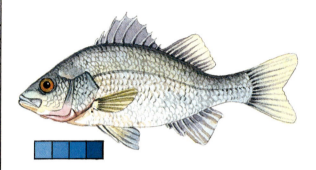

The white perch *(above)* is one of the few resident species of the bay. The perch tolerates the fluctuating salinity but spawns in freshwater tributaries. Other estuarine species include the anchovies and silversides, killifish, blennies, gobies, rockfish, and the fancifully named hogchoker.

Salinity in parts per thousand

0-5   5-10   10-15   15-18   18-25   25-30

Autumn salinity levels in the Chesapeake Bay. One of the most important factors determining the variety and amount of life in the bay is the saltiness of the water. The water is too salty for most freshwater species and not salty enough for most saltwater species. The level of salinity varies during the year: It drops in the spring or after a rainstorm because of freshwater runoff and rises during the drier months of summer and autumn.

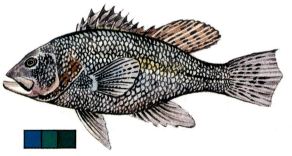

Black sea bass *(above)* are marine, or ocean-dwelling, fish. They spawn in the ocean in spring, then swim into the salty lower bay to feed on small fish and invertebrates during the summer months. Other species that seasonally migrate into the bay include sandbar shark and cow-nosed ray, shad, bluefish, drum, flounder, and lizardfish.

Do Oil Spills Damage the Environment?

Accidental oil spills befoul beaches and threaten a coastline's delicate ecological balance. The 1989 spill in Alaska's Prince William Sound, for instance, contaminated one of the world's most pristine wildlife habitats with 11 million barrels of crude oil. As a result, nearly 3,000 otters and some 300,000 birds died, along with countless mollusks and plants sensitive to even the slightest presence of toxic oil.

Although nature heals itself and the most visible scars oil leaves on coastlines disappear within five years, in the short term an oil spill constitutes a disaster, and there is much debate about the long-term ecological damage to areas that have been polluted by oil. Scientists are particularly concerned about potential genetic damage to fish, disruptions of the food chain, and other effects still poorly understood.

Animals with little mobility, such as mussels and starfish, are most exposed to the toxins. Because they are food for other species, their death disrupts the food chain, threatening even large mammals like bears, which feed on fish.

Sea otters suffer greatly during oil spills. Many freeze to death, since oil destroys the insulating properties of their fur. Others may get sick from ingesting oil while trying to clean themselves. Seals and other animals that rely on an internal layer of fat to keep warm suffer less.

Fate of an oil slick

When a tanker leaks oil into the water, the resulting slick spreads out into a broad, thin film. As much as half of this film quickly evaporates, removing most of the highly volatile toxic components. Wind and water break up what is left into small particles, which either disperse or combine with water, in a process called emulsification, to form viscous clumps, called mousse. Photooxidation—the effect of sunlight on oil molecules—chemically disintegrates some of the mousse. After a month a significant fraction of the oil begins to biodegrade, or break down, as marine organisms and bacteria consume it. Finally, what is left of the mousse—about 20 percent—forms into tar balls, which remain in the environment.

Evaporation

Wind direction

Photooxidation

Direction of spill

Mousse

Sedimentation

Biodegradation

When a tanker begins to leak, additional ships are dispatched to unload the oil that has not yet spilled. To contain the oil already released, cleanup crews surround the slick with floating booms *(red)*, then vacuum up the oil with machines called skimmers.

Birds are the most visible victims in an oil spill. Some, like the murres shown here, become covered in oil, which makes them unable to float or stay warm. Others, like eagles, get sick from feeding on contaminated fish. Since bird eggs are sensitive to even trace amounts of oil, a contaminated bird returning to a nest can destroy an entire clutch of eggs.

Oil and feathers. Fine networks of interlocking hooks *(above, left)* make bird feathers highly insulating and watertight. Oil prevents these hooks from linking together, leaving contaminated birds exposed and likely to die.

What Is Biological Accumulation?

When pesticides came into wide use in the 1950s, the long-term effects of the chemicals were not yet known. Around the world, however, fish-eating water birds on lakes and marshes treated with pesticides to control mosquitoes began dying in droves. Autopsies revealed that the dead birds contained pesticides in concentrations many times greater than the density of the chemicals carried in the water.

How those poisons had built up to such high levels was uncovered by examining the food chain of the grebe *(below)*. Toxic chemicals intro-duced into a body of water are absorbed by phy-toplankton, then passed on to the fish and other creatures that feed on it. These herbivores transmit the toxins to the grebe and to carnivo-rous fish that prey on them. At each link in the food chain, the chemicals are increasingly con-centrated; at the top of the chain, the amount of poison may be enough to kill the grebe or large fish or impair their ability to reproduce. This phe-nomenon—whereby a substance accumulates in ever greater densities as it moves up the food chain—is known as biological accumulation.

Grebe

Plankton

Aquatic plants

Carnivorous fish

Herbivorous fish

Glossary

Acid rain: Rainwater that contains strong concentrations of acid-forming chemicals.

Adaptive insulation: The development of extra protection from the cold—such as an extra layer of fur or an internal layer of fat—that helps animals survive in frigid climates.

Amphibian: Any of a class of animals, including frogs and salamanders, most of which are born with gills but develop lungs.

Amphibious animals: Animals that can live on land and in water.

Anaerobic soil: Soil that contains little or no oxygen.

Anthropoids: Members of the superfamily that includes humans and the great apes—gibbons, orangutans, chimpanzees, and gorillas, as well as their evolutionary ancestors.

Archipelago: A long chain of islands, often volcanic in origin, such as the Galápagos.

Artiodactyls: Hoofed animals that have an even number of toes, such as the giraffe, antelope, buffalo, and sheep.

Basidia: Reproductive cells of certain higher fungi that bear spores. In mushrooms they are located in the gills.

Biological accumulation: The process in which concentrations of chemicals such as pesticides build up in the bodies of animals that consume previously contaminated plants and animals.

Biome: A broad area of terrestrial plant and animal communities characterized by its soil and climate. Earth contains nine major biomes: tropical rain forest, savanna, grassland, desert, chaparral, deciduous forest, coniferous forest, taiga, and tundra.

Biogeographic regions: Distinct geographical areas of the world, characterized by their climates and the plants and animals that live in them. Earth's regions include: the Nearctic Region of North America, the Palearctic Region of Eurasia and North Africa, the Oriental Region of southern and eastern Asia, the Neotropical Region of the Caribbean and Central and South America, the Ethiopian Region of Africa, and the Australian Region.

Biosphere: The combined regions of the earth, including the land, oceans, and atmosphere that sustain all life.

Breathing root: A type of root found in species of mangrove trees that extends above the ground to take in air.

Carnivore: An animal or plant that eats animals; also called a **secondary consumer.**

Carrion: The flesh of dead animals.

Cauliflorous trees: Trees that sprout flowers, leaves, and fruit directly from their trunks, rather than from new branches.

Chlorophyll: The green coloring matter in plants that allows them to use sunlight to manufacture starch through the process of photosynthesis.

Classification: The orderly arrangement of plants and animals intended to show evolutionary relationships. From the most general to the most specific the levels are: kingdom, division (plants) or phylum (animals), class, order, family, genus, and species. Intermediate levels are termed sub- or super- as in suborder and superfamily. Examples are:

   White oak: Plantae, Anthophyta, Dicotyledoneae, Fagales, Fagaceae, *Quercus alba*

   Chimpanzee: Animales, Chordata, Mammalia, Primates, Pongidae, *Pan troglodytes*

Coevolution: The process in which two or more interacting species evolve interdependently.

Cold-blooded animal: An animal without the ability to regulate its body temperature. Such animals must absorb heat from their surroundings to keep warm, as do fish, amphibians, and reptiles.

Colonization: The process by which species of plants and animals move, or are moved, into a new area and survive to reproduce.

Crustacean: Any of a class of invertebrate animals, including barnacles, crabs, shrimp, and lobsters, that are mainly aquatic and characterized by hard but flexible outer skeletons, possess lungs, and have two pairs of antennae.

Desert: An area that receives less than 10 inches of rainfall per year. Deserts are home only to hardy plants and animals.

Differentiation: The development of characteristics that allow a species to function more efficiently in its habitat.

Dinosaurs: A subclass or superorder of reptiles, many of which were extremely large, that lived on earth for some 160 million years; they became extinct about 65 million years ago.

DNA (Deoxyribonucleic acid): A molecule found in all life on earth that is responsible for storing genetic information.

Drought: A prolonged period in which little or no rain falls.

Echolocation: The process by which some insectivorous bats and cave-roosting birds navigate by sending out high-frequency sound waves and gauging distances by their echoes.

Ecology: The study of how living things interact with each other and their environment.

Ecosystem: A major subdivision of the biosphere containing a number of interacting plant and animal species and their physical environment.

Embryo: An immature stage of an animal in its mother's womb or a plant in a seed that cannot survive on its own.

Endocarp: The inner layer of a plant ovary wall.

Endosperm: A section of a plant embryo containing nutrient tissue used by the embryo for food.

Epiphytes: Plants that live above the ground on trees, absorbing moisture and nutrients directly from the air. As they grow, they may send roots to the ground and strangle the trees on which they live; such plants are also called stranglers.

Erosion: The process in which wind and rain eat away the topsoil from an area.

Evaporation: The process by which a liquid turns into a vapor.

Evolution: The process in which species develop new physical characteristics through genetic alteration, ultimately becoming new species.

Exocarp: The outer layer of a plant ovary wall.

Food chain: A pattern of feeding, in which species in an ecosystem consume other species. Producer plants, which use sunlight to produce starch, are at the bottom of the food chain. Herbivorous animals are the next step up, while carnivorous animals make up the top end of the chain.

Fungi: A group of plants, including mushrooms and molds, that lack chlorophyll and therefore cannot make their own food. Many are parasites living on other plants or animals.

Genes: The fundamental units of heredity in plants and animals.

Germination: The process in which a seed begins to grow.

Gills: In plants, the dark platelike structures beneath the cap of a mushroom that radiate outward from the stem and bear spores. In fish and aquatic invertebrates, they are the blood-filled organs by which oxygen is absorbed from water.

Grassland: An expanse of land covered predominantly by grass, such as the **prairies** of North America, the **savannas** of Africa, the **steppes** of Eurasia, and the **pampas** of South America.

Guano: The droppings of birds, bats, and seals that accumulate in caves and seabird colonies. Guano is rich in nitrogen and phosphates and is useful as a natural plant fertilizer. Insects and other animals low on the food chain may consume guano.

Habitat: The particular environment in which a plant or an animal lives.

Herbivore: An animal that eats only plants.

Hibernation: A practice common to many species in which an animal's metabolism slows down for sleep through the winter.

Humidity: A measure of the amount of water vapor in the air.

Hyphae: The thready filaments that form the mycelium of a fungus.

Ice age: One of several periods in earth's history during which much of the planet's surface was covered with ice.

Incubation: The process in which animal eggs are kept warm, either by a parent or by the sun, or by decaying vegetation or moderate volcanic heat in a buried nest. The warmth allows the embryo in each egg to mature until it is ready to hatch.

Insect: Any of a class of small animals, such as bees, ants, and beetles, characterized by hard outer skeletons, three distinct body segments, and three pairs of legs.

Isopod: Any of an order of small crustaceans, each of which posesses from two to eight pairs of legs.

Ivory: The material that makes up the tusks of elephants, walruses, and narwhal whales.

Keystone species: A species that has a great influence on the other plants and animals in its habitat.

Larva: The immature, wormlike form of a newly hatched insect and some other invertebrates; also referred to as a grub.

Lenticels: Small openings on the roots, stems, and leaves of vascular plants that allow an exchange of gases.

Mammals: A class of warm-blooded animals characterized by their nursing of their young.

Marsupials: An order of mammals, including kangaroos and opposums, whose young are born at an early developmental stage; the young develop further in the mother's abdominal pouch.

Mesocarp: The middle layer of a plant ovary wall.

Metabolism: All of the biochemical processes necessary to sustain life; they include respiration, digestion of food, and buildup of new body tissues.

Migration: The periodic movement of groups of animals, such as flocks of birds or schools of fish, to a location with a more favorable climate or abundant food.

Molting: The process by which many animals periodically shed and regrow their skin, hair, or feathers.

Mycelium: The part of the mushroom that lies beneath the ground.

Mycorrhiza: A type of symbiosis in which a mushroom or other fungus exchanges nourishment with the roots of a living plant.

Niche: A role played by a species within an ecosystem.

Ovule: The female element of the reproductive systems of flowering plants that receives pollen to produce a seed.

Parasite: An animal that lives in or on another animal, called a host, and obtains food from it. The presence of a parasite usually harms, and often kills, its host.

Perennial plant: Any plant that persists year-round or regrows every year in an area.

Pesticides: Chemicals used to control harmful insects or plants.

pH: A scale for the acidity of fluids, running from 0—the most acidic—to 14—the least acidic.

Photooxidation: The chemical process in which oil breaks down when exposed to sunlight.

Photosynthesis: The chemical process in which chlorophyll-bearing plants use energy from sunlight to manufacture starches.

Placental mammals: The name given to mammals whose young develop inside the mother's womb.

Plankton: Minute plants, called phytoplankton, and animals, called zooplankton, that live in water.

Pneumatophores: Above-ground or water extensions of the roots of certain trees that grow in water or poorly drained soils and absorb oxygen, channeling it to the true roots.

Pollen: Tiny male spores produced by plants that are necessary for the fertilization of seeds.

Pollination: The transfer of pollen from stamen to ovule; a process necessary for the fertilization of plant seeds.

Predator: Any animal that hunts and eats other animals.

Prehensile tail: A tail sometimes found in primates and marsupials that is specially adapted for grasping and support.

Primates: The order of mammals that includes apes, monkeys, and humans.

Prosimian: Any member of a suborder of primates that includes the early ancestors of present-day monkeys as well as some primitive living forms, such as lemurs.

Protozoa: Single-celled creatures like paramecia and amoebas that along with plants and animals, make up the three major groups of living things.

Rain forest: A forest that receives at least 80 inches of rain annually.

Raptor: Any bird of prey, such as the hawk or owl.

Reptile: Any member of a class of cold-blooded animals that have, among other common traits, skin made of scales or hard plates.

Rodent: Any member of an order of mammals, including the mouse, rat, and beaver or muskrat, commonly characterized by a set of prominent front teeth used for gnawing and large hind molars for grinding.

Rotifers: A phylum of microscopic aquatic animals near the bottom of aquatic food chains.

Runner: An offshoot of the stem of some plants that gives rise to new plants as it extends along the ground.

Salt glands: Special structures in the roots of some trees or the heads of some marine reptiles and birds that filter out and excrete salt from water or blood.

Scavengers: Animals that live off the flesh of dead animals.

Slash-and-burn agriculture: A method used in the tropics in which all trees are cut down and burned to clear for planting.

Soil horizon: The term given to a layer of soil. The three soil horizons from top to bottom are topsoil, subsoil, and bedrock.

Spore: The reproductive unit of ferns, mosses, mushrooms, and other fungi, similar to seeds in other types of plants.

Stamen: The male portion of a flower that produces the pollen necessary for the fertilization of seeds.

Stomata: The pores in plant leaves through which carbon dioxide is absorbed and oxygen and water vapor released. Similar respiratory openings are on the bodies of insects.

Symbiosis: A close relationship between two species in which both species benefit from the relationship.

Transpiration: The process in which plants emit water vapor.

Troglophiles: Animals, such as bats, that live in caves but venture outside to hunt for food.

Vascular plant: Any higher plant that circulates liquid and food throughout its body.

Warm-blooded animal: An animal that maintains a constant body temperature, independent of the surrounding temperature.

Wetlands: Areas, such as bogs, swamps, or marshes, where the soil is either very moist or covered with water.

Index